DESIGNING CITIES WITH PEOPLE, NOT JUST FOR THEM

emerald PUBLISHING
ice Publishing

DESIGNING CITIES WITH PEOPLE, NOT JUST FOR THEM: THE POWER OF COLLABORATIVE URBAN FUTURES

Practical Insights for Planners, Leaders and Citizens

BY

DR ARCHANGELO D ALEGADO
PhD MPSA MTSS MEP

DR REC E EGUIA
PhD MPSA MM

DR ALAN B ALEJANDRINO
PhD MPSA MATM

Published by Emerald Publishing Limited, Floor 5, Northspring, 21-23 Wellington Street, Leeds LS1 4DL.

ICE Publishing is an imprint of Emerald Publishing Limited

Other ICE Publishing titles:

Sustainable and Resilient Cities: Insights into Urban Planning in US Cities and Communities
Cesar A. Poveda ISBN 978-1-80686-688-5
Digital Twins for Smart Cities: Conceptualisation, Challenges and Practices
Li Wan, Timea Nochta, Junqing Tang and Jennifer Schooling
ISBN 978-0-7277-6600-7

A catalogue record for this book is available from the British Library
ISBN 978-1-83742-369-9

Cover photo: Shutterstock.com/g/zhuxiaozhu

Commissioning Editor: Michael Fenton
Content Development Editor: Ryan Molyneux
Books Production Lead: Emma Sudderick

Typeset by KnowledgeWorks Global Limited
Index created by David Gaskell

Printed and bound by CPI Group (UK) Ltd, Croydon, CR0 4YY

CONTENTS

PREFACE

"Cities are not just where we live. They are what we make of our collective hopes, and sometimes, what we neglect in our silence. This book is about making sure our voices shape them."

When I began this journey researching how Digos City could transform into the 'city of choice' by 2040, I thought of it merely as an academic exercise. But as the research progressed, it became clear: this is no longer just about Digos. This is about the kind of cities we all want to live in and how we can get there together.

In too many places, city planning remains a top-down affair. Well-meaning professionals and politicians make plans *for* people, often without inviting them into the conversation in a meaningful way. Yet, time and again–in Helsinki, Seoul, Barcelona, Singapore and beyond–evidence shows that when citizens are truly part of the process, cities become more resilient, inclusive and innovative.

This book draws on my doctoral research on Digos City, a city in Southeastern Mindanao poised at the threshold of great change, grappling with challenges familiar to many cities across the Philippines and the world: rapid urbanisation, weak participation, fragmented planning, pressing ecological and social demands, and corruption. It also weaves in lessons from cities around the globe that have succeeded–or stumbled–in their own urban transformation journeys.

Together with my mentors, our goal here is simple: to help planners, leaders and everyday citizens understand how collaborative foresight–the art of planning cities *with* people, not just *for* them–can shape better urban futures.

Whether you are a city planner facing political pressures, a community leader trying to engage residents, or simply a citizen who cares about the place you call home, this book offers stories, insights and practical tools to help you become an active participant in building the future of your city.

Throughout, we state each lesson as a general principle first, then show how it plays out in Digos, so that readers in any city can adapt it.

Let's begin.

Archangelo D Alegado PhD MPSA MTSS MEP, 2026

ACKNOWLEDGEMENTS

This book is the product of not just our research, but of countless conversations, collaborations and quiet encouragement along the way.

We are deeply grateful to the dissertation committee members for their wisdom, patience and rigorous guidance that sharpened both our ideas and our resolve. To the participants of this study–the citizens, leaders and stakeholders of Digos City–thank you for sharing your stories, your hopes and your candour. This work is as much yours as it is ours.

To our mentors, colleagues and students at Cor Jesu College and the National Police College, you challenged us to see beyond our own assumptions and reminded us of the transformative power of learning together.

Most of all, to our families, whose unwavering support, love and understanding carried us through long nights and early mornings, thank you for believing in this journey and in us.

May this book honour all of you, and may it inspire others to continue building cities worthy of their people.

Designing Cities with People, Not Just for Them: The Power of Collaborative Urban Futures

Archangelo D Alegado, Rec E Eguia and Alan B Alejandrino
ISBN 978-1-83742-369-9
https://doi.org/10.1108/978-1-83742-366-820261002

INTRODUCTION

> *"The future of a city is neither guaranteed nor accidental—it is built in the choices we make and the people we include."*

We stand at a crossroads in how we shape the cities we live in. Cities are growing faster than ever before–not just in size, but in complexity. More than half of the world's population now lives in urban areas. By 2050, this number will rise to two-thirds. These figures are more than just statistics. They are a reminder that the quality of our urban life is one of the defining challenges and opportunities of our time.

On one hand, cities are extraordinary engines of growth, creativity and innovation. They concentrate talent, offer economic opportunities and bring people from different walks of life together. On the other hand, they are also where inequality festers, where poor planning creates congestion and chaos and where the disconnect between government and citizens is most visible.

The Philippine experience embodies this tension perfectly. The country's urban population is rising at a rapid pace, from around 51% in 2015 to 54% in 2020. It is on track to surpass 65% by 2050. Cities like Digos are not exempt from this trend.

And yet, in too many cases, the way we plan our cities remains stuck in the past: top-down, politically driven and too often blind to the voices of the very people who live in these spaces. Plans are made for people, not with them. Decisions are shaped by expediency rather than foresight. Citizens, when they are engaged at all, are brought in only as a formality.

This disconnect has consequences. Roads crumble, drainage systems fail, economic opportunities bypass the marginalised and trust in governance erodes. As urban populations swell, so too do the challenges. If they remain unaddressed, the very vitality of our cities is at stake.

THE CASE OF DIGOS CITY

Digos City, in Southeastern Mindanao, is a microcosm of these dynamics. On paper, it has made impressive strides in recent years: *Most Improved Component City* (2021), *Second Most Competitive City in Innovation in Davao Region* (2022) *and Multiple Seal of Good Local Governance Awards.*

These accolades are a testament to its leadership and its potential. But beneath the surface lies a different story, one that mirrors many cities across the Philippines and beyond, where

- political interference delays projects and undermines transparency
- citizens have little awareness of, or say in, decisions that affect their daily lives
- infrastructure lags population growth, creating traffic, pollution and risk

- marginalised communities remain excluded from the vision of progress.

It is a city at a crossroads: caught between its promise and its peril.

In 2019, its leadership articulated an ambitious vision: to transform Digos into the city of choice, a city people choose to live in, work in, invest in and visit. But a vision is only as good as its execution—and execution, in turn, depends on who gets to participate in shaping it.

Digos is not alone in this struggle. Cities everywhere face similar tensions: between ambition and reality, growth and equity, tradition and innovation.

WHY THIS BOOK?

This book is a response to these challenges and to an urgent question: *How can cities be designed with people, not just for them?*

Over the course of the author's doctoral research on Digos City, together with mentors, they explored how collaborative foresight–the ability of communities, governments and stakeholders to imagine and plan their future together–can help cities escape the trap of top-down planning and unlock more inclusive, resilient futures.

Their findings were both sobering and hopeful: sobering, because the obstacles are real: power imbalances, institutional inertia, corruption, lack of trust and resource constraints; hopeful, because these challenges can be overcome when we engage citizens meaningfully, build trust and promote shared ownership of urban futures.

This book translates those insights–rooted in research but written for practitioners–into practical lessons, stories and tools for planners, leaders and citizens alike.

We begin here, at the threshold of decision, in the space between promise and peril, where the future of our cities hangs in the balance, waiting for us to choose.

HOW TO READ THIS BOOK

This book is written for city leaders, planners and communities everywhere, not just for Digos. Each chapter follows a simple pattern, beginning with a principle that applies to cities broadly. The principles are then illustrated with a global or comparative case and finally it is shown how it plays out in Digos. In this way, readers can see both the universal lesson and its practical application in a real city context. By keeping this rhythm throughout the book, the insights are transferable and useful whether for endeavours in a large metropolis, a growing secondary city or a small community taking its first steps in participatory planning.

Figure 1 below explains where the book's ideas come from. While this book is designed to be read like a practical guide, it is grounded in a clear research logic. The ideas, stories and tools presented in the chapters that follow were not assembled at random; they emerged from a structured process of analysing Digos City's context, imagining alternative futures and testing what really matters for urban transformation.

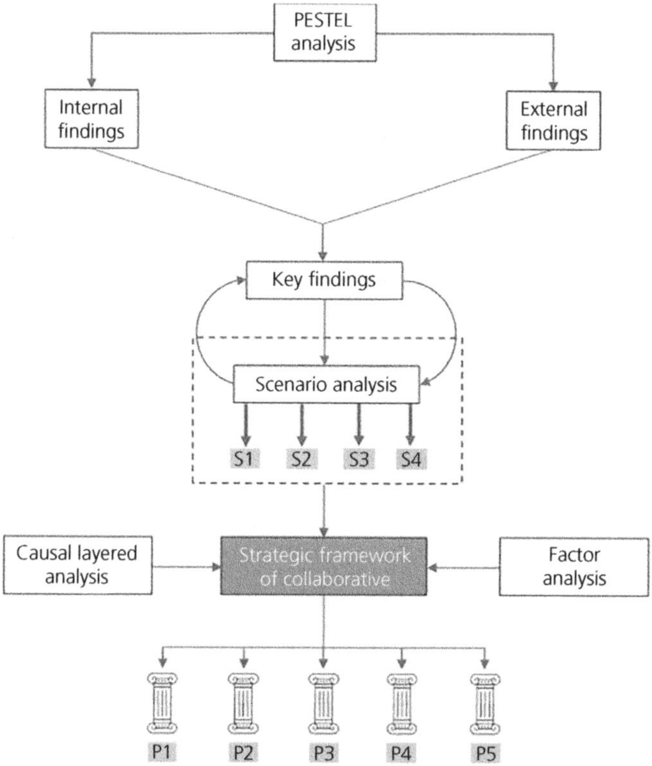

Figure 1 Methodological framework for concurrent and sequential synthesis

Part I

UNDERSTANDING THE LANDSCAPE

Designing Cities with People, Not Just for Them:
The Power of Collaborative Urban Futures

Archangelo D Alegado, Rec E Eguia and Alan B Alejandrino
ISBN 978-1-83742-369-9
https://doi.org/10.1108/978-1-83742-366-820261005

1

WHY CITIES MATTER – AND WHY THEY STRUGGLE

> *"We shape our cities, and thereafter they shape us."*
> *– Winston Churchill*

Cities are the stage upon which the drama of modern life unfolds. They are where economies grow, cultures collide and converge, and innovations take root. In many ways, cities are humanity's greatest invention – places where the sum becomes greater than its parts.

But they are also where inequality is most visible, where poor planning leaves scars, and where the cracks in our systems show first. For every vibrant, resilient city, there are dozens struggling under the weight of mismanagement, exclusion and unbridled growth.

Why do cities matter so much? And why do they so often fall short of their promise?

THE GLOBAL PUSH FOR SUSTAINABLE CITIES

In recent decades, the international community has recognised just how critical cities are to our shared future.

In 2015, the United Nations adopted the Sustainable Development Goals (SDGs), a blueprint to achieve a better and more sustainable future for all. Among them, SDG11 specifically focuses on cities, urging nations to 'make cities and human settlements inclusive, safe, resilient, and sustainable'.

One year later, the New Urban Agenda, agreed upon at the Habitat III conference in Quito, Ecuador, built on this momentum. It calls on cities around the world to lead the charge toward sustainability that balances economic growth, environmental protection and social equity.

These frameworks are ambitious but necessary. They recognise that the battle for sustainability will largely be won or lost in cities–the engines of growth, but also the biggest consumers of resources and producers of waste.

Cities like Helsinki in Finland, Barcelona in Spain, Seoul in South Korea and Singapore, have already embraced this challenge, implementing long-term plans anchored in sustainability, public participation and innovation. Helsinki has pursued a 'functional city' strategy that puts people and services at the centre. Barcelona chased an 'entrepreneurial city' strategy. Seoul, through its 2030 Plan, has reimagined its streets, its governance and its public spaces with citizens as co-creators of change. And Singapore, a city-state, established a 'smart nation' approach.

These examples show what is possible when vision meets action. But they also highlight a sobering reality: not all cities have the same resources, capacity or political will to follow suit.

WHY LOCAL REALITIES OFTEN LAG GLOBAL AMBITIONS

If global goals are so clear, why do so many cities fall short? The answer lies in the gap between vision and reality–between what is imagined in global forums and what is lived on the ground.

For many developing cities, several obstacles stand in the way. *Political cycles* trigger leaders to prioritise quick wins to satisfy short-term mandates, leaving little room for long-term planning. *Weak institutions* are without strong governance and transparency; resources are mismanaged and opportunities lost. *Inequality and exclusion* leave marginalised voices out of planning, leading to solutions that favour the few. *Resource constraints* mean limited funding, technical expertise and reliable data, making it harder to implement ambitious plans.

In the Philippines, these challenges are pronounced. Local governments remain heavily dependent on national government transfers. Planning horizons are often tied to political terms rather than generational visions. Participation, when it happens, is more symbolic than substantive.

This is not just a governance problem–it is a structural one. Yet, it is precisely in these contexts that meaningful, collaborative foresight becomes even more essential.

WHAT THIS CASE REVEALS ABOUT THE PHILIPPINE URBANISATION STORY (WITH DIGOS AS AN ILLUSTRATION)

The story of Philippine urbanisation is not only about its largest metropolitan centres. While Metro Manila, Cebu and Davao dominate the headlines, it is the secondary cities that carry the next wave of growth. These cities absorb migration, provide regional services and anchor local economies. Yet, they often face the same challenges as larger cities–weak infrastructure, limited fiscal autonomy and fragile planning–without the same resources to manage them. The principle is clear: the long-term trajectory of national development depends heavily on how these secondary cities are able to balance growth, equity and sustainability.

Digos is one of these cities. As a growing secondary hub in Mindanao, it reflects both the opportunities and the vulnerabilities of Philippine urbanisation. Its population growth, service demands and economic shifts mirror the pressures experienced by many similar cities across the archipelago. When we look at Digos, we see a living example of how these forces converge and play out in real time.

In Digos, this appears in the mismatch between urban expansion and service provision. Drainage and road systems lag behind residential growth, while social services remain under strain from limited budgets. Planning often becomes reactive, shaped more by short-term projects than by a sustained, long-horizon vision. At the same time, community voices struggle to find traction in decision making, reinforcing the sense that urbanisation is happening to people rather than with them. These patterns are not unique to Digos–they are the lived reality of many mid-sized Philippine cities trying to keep pace with rapid change.

For other secondary cities, the lesson is straightforward: urbanisation cannot be left to inertia or to the political cycle. The cue is to localise global commitments like SDG11–sustainable cities and communities–while insulating long-term planning from election-driven shifts. In practice, this means embedding foresight, codifying city development agendas that stretch across terms, and investing early in infrastructure that shapes growth rather than merely reacts to it. Digos illustrates what is at stake, but the principle is one that resonates across the Philippine urban system and in many developing countries facing the same pressures.

WHY THIS MATTERS

The story of Digos reflects the struggle of many cities in the Philippines and around the world: caught between global aspirations and local constraints, between promise and peril.

It is precisely because of these struggles that the lessons from Digos–and the strategies discussed in this book–are so important. Why this matters is not only for Digos, but also for all secondary cities grappling with fast growth and fragile planning. The principles set out here are transferable, and Digos serves as one illustration among many.

If we want our cities to be resilient, inclusive and sustainable, we cannot continue with business-as-usual. We must recognise that cities are not just physical spaces, they are social contracts, shaped by the people who inhabit them. And the future of our cities depends on whether those people are invited to the table, heard and empowered.

In the next chapters, we will look at the possible futures for Digos City by 2040, exploring scenarios that illustrate both the risks of inaction and the opportunities of collaborative foresight.

Archangelo D Alegado, Rec E Eguia and Alan B Alejandrino
ISBN 978-1-83742-369-9
https://doi.org/10.1108/978-1-83742-366-820261006

2

THE STATE OF DIGOS CITY

"A city is not just measured by its skyline, but by the lives it shapes–and those it leaves behind."

SITUATING DIGOS IN URBAN TYPOLOGIES

Cities are not all the same. Their type–the role they play within national and regional systems–shapes both their opportunities and their risks. Typologies provide a comparative lens: they allow us to see where a city fits in relation to others and why its struggles to take a particular form.

Globally, four broad patterns emerge. Metropolitan hubs, such as Manila or Jakarta, concentrate political power, resources and services but suffer from congestion, inequality and overstretched infrastructure. Peri-urban expansion zones, typically found on the fast-growing edges of larger cities, face uncoordinated growth, weak governance and fragile public services. Polycentric systems, such as the Randstad in the Netherlands, distribute development across multiple urban centres, creating balance and resilience through interconnection. Finally, secondary cities–smaller but significant hubs outside primate capitals–serve as anchors for balanced regional development. They often lack visibility in national policy, yet their contribution to equity and sustainability is critical (Roberts, 2014; UN-Habitat, 2020).

Digos is a secondary city within the Philippine urban hierarchy. It is not a megacity like Metro Manila, but neither is it peripheral. Its position between Davao City and its surrounding municipalities makes it a hinge for growth in Davao del Sur. As a secondary city, Digos carries both promise and vulnerability. It can absorb pressures that would otherwise intensify in metropolitan centres but, if unmanaged, it could also replicate the fragility of peri-urban sprawl. Recognising Digos as a secondary city clarifies its strategic role–balancing national urbanisation while building resilience through planned density, regional connectivity and inclusive governance.

Seen through the lens of urban typologies, Digos stands out as a secondary city whose trajectory will be defined not by sheer scale, like Manila, but by how it manages governance, participation and infrastructure within its regional role. The following sections show how these dynamics play out on the ground–making Digos a microcosm of both the challenges and possibilities faced by many growing cities worldwide.

DIGOS AS A MICROCOSM OF URBAN DEVELOPMENT

Digos City sits at the heart of the Philippines urban transformation story. Neither a metropolis like Davao City nor a sleepy rural town, Digos represents the reality faced by many medium-sized cities in developing countries: poised for growth but constrained by structural, social and political barriers.

It is a city where rapid urbanisation collides with fragile infrastructure; where ambition shines brightly on paper but falters in execution; and where awards and accolades coexist with unaddressed inequalities and neglected voices.

Digos embodies the tension between what cities aspire to become and what they actually are. That is precisely why its story matters–because it mirrors the struggles and hopes of countless other cities striving to balance growth, inclusion and sustainability.

ACHIEVEMENTS, RECOGNITIONS AND GROWTH

On the surface, Digos City has much to be proud of. Over the past few years, it has earned distinctions that many other cities in the region envy. Its population growth–among the highest in the Davao region at 2.31% annually–is both a sign of its attractiveness and a source of mounting pressure. As people continue to move in, the city's economy expands, businesses emerge and opportunities increase.

The city's vision to become the 'city of choice' or a place where people choose to live, work, invest and visit, speaks to its aspirations to rise beyond mediocrity and become a model of inclusive urban growth.

These achievements, however, tell only part of the story.

THE SYSTEMIC AND HUMAN CHALLENGES BENEATH THE SURFACE

Behind the accolades lie systemic issues that threaten to undermine the city's progress. Digos, like many growing cities, faces a set of challenges–some visible, others hidden–that reflect the deep contradictions of urban development in the Philippines.

Governance gaps

Governance in Digos City still bears the marks of old habits–habits that slow progress, frustrate citizens and erode trust. Political interference continues to distort priorities, with decisions sometimes driven more by who holds influence than by what the community truly needs. This dynamic has skewed resources and delayed important projects, feeding a sense of disillusionment among citizens who feel sidelined from decisions that shape their daily lives.

Administrative lapses have compounded the problem. Public works are delayed far too often, not just because of funding gaps but because of weak coordination and poor follow-through. The absence of an updated executive-legislative agenda leaves the city without a shared long-term vision, leaving departments to react to problems piecemeal instead of working toward strategic goals.

At a deeper level, there remains a troubling disconnect between government institutions and the people they serve. Too few residents know about the city's development plans and even fewer feel they have a real say in shaping them. Formal participation mechanisms exist on paper, but they often fail to reach marginalised groups or to inspire genuine dialogue. Too many citizens see their role as passive–waiting for announcements (e.g. public hearings and consultation) rather than being invited to co-create solutions.

These gaps are more than just bureaucratic missteps; they reflect a larger pattern of governance that has yet to fully embrace transparency, accountability and inclusiveness. Without intentional change, the city risks falling into a cycle of 'hand-me-down' governance–where decisions trickle down from above rather than emerging from collaboration with the people most affected.

Bridging these gaps requires more than technical fixes. It calls for a shift in mindset and practice: leadership that welcomes scrutiny, institutions that empower citizens and systems that reward integrity over expediency. Until governance becomes a two-way street, where both leaders and communities feel responsible for outcomes, transformation will remain a distant vision rather than a shared reality.

When planning is decoupled from a shared executive-legislative agenda, cities drift toward reactive fixes. Codifying a cross-term roadmap stabilises priorities beyond political cycles and creates predictability for communities and investors alike.

Weak public participation

For many residents of Digos City, participation in governance feels more like an afterthought than a right. While formal mechanisms for engagement technically exist, such as development councils, special committees and barangay assemblies, they are too often underused or treated as box-checking exercises rather than genuine opportunities for collaboration.

This has left citizens feeling more like passive observers than active stakeholders in shaping their city's future. Many remain unaware not only of the city's plans but also of their own role and rights in influencing them. Those who do attend consultations frequently encounter decisions that already seem predetermined, fuelling scepticism about whether their voices truly matter.

Marginalised groups–including informal settlers, relocated families and micro-entrepreneurs–are particularly excluded, not just because of weak outreach but because the language, timing and format of engagement often fail to meet them where they are. Meetings are held during working hours or conducted in ways that feel inaccessible and intimidating.

Participation, when it does occur, tends to hover at the lower rungs of Arnstein's Ladder (Arnstein, 1969)–closer to 'informing' or 'consultation' than to partnership or citizen control. As a result, public input rarely translates into shared ownership of decisions or sustained civic engagement.

If participation is to strengthen, it must move beyond tokenism. Citizens need to see that their contributions shape real outcomes–that their knowledge of their neighbourhoods, their struggles and their aspirations is valued. Only then will they begin to step out of the role of spectators and into the role of co-creators of the city they inhabit.

Tokenistic participation is not unique to Digos; it recurs wherever engagement is last-minute and symbolic. Embedding participation at the start of planning, with visible 'you said, we did' feedback loops, is a simple but powerful way to build credibility across contexts.

Fragile infrastructure

As Digos City expands, its physical and digital infrastructure struggles to keep up. The road network, already strained by increasing vehicle use and encroachment, contributes to congestion and accessibility challenges across neighbourhoods. Drainage systems, often inadequate or poorly maintained, leave many areas vulnerable to flooding, especially during heavy rains–a recurring problem that undermines both safety and economic activity.

Public amenities such as parks, markets and community spaces are limited in both number and quality, failing to meet the needs of a growing and diversifying population. Basic facilities that underpin public health and environmental sustainability–like a proper wastewater treatment plant–remain absent, leaving the city reliant on outdated and often harmful practices.

The city's digital infrastructure is equally underdeveloped. E-governance systems that could streamline services, improve transparency and bring government closer to citizens are still in their infancy, making interactions with local offices cumbersome and inefficient. Without these modern systems, both citizens and businesses encounter unnecessary friction, eroding confidence in the city's ability to deliver a more efficient and responsive urban experience.

In many ways, the fragility of Digos's infrastructure mirrors the fragility of its governance–reactive rather than proactive, incremental rather than transformative. Left unaddressed, these weaknesses not only compromise the city's quality of life today but also jeopardise its ability to fulfill its vision of becoming a resilient, people-centred city of choice in the future.

Infrastructure weakness is a common challenge in secondary cities that expand faster than their fiscal base. The principle is to link investment decisions to long-term resilience–prioritising assets like drainage and transport that shape the city's ability to absorb growth.

Lack of strategic roadmap

Perhaps the most critical gap in Digos City's journey toward becoming the city of choice is the absence of a clear, unifying strategic roadmap. Without a long-term development framework to anchor decisions and investments, the city risks drifting from one administration to the next, each with its own set of priorities, often disconnected from the bigger picture.

In place of a cohesive vision, planning tends to be fragmented and reactive–focused on short-term fixes rather than sustainable solutions. Projects are too often shaped by immediate political pressures or narrow interests, rather than by a shared understanding of where the city needs to be in 10, 20 or even 30 years. This not only wastes resources but also creates a sense of inconsistency and unpredictability, which undermines trust in government and discourages broader stakeholder investment.

The lack of a strategic roadmap also means missed opportunities to coordinate across sectors–linking infrastructure, economic development, environmental sustainability and social inclusion into a single coherent strategy. Without such alignment, even well-intentioned initiatives risk working at cross-purposes or failing to achieve meaningful impact.

A strategic roadmap does more than chart a path forward–it helps leaders, citizens and partners see themselves as part of a collective effort to shape the city's future. Until Digos develops and adopts such a plan, its vision of becoming a resilient, inclusive and thriving city of choice will remain vulnerable to the shifting winds of politics and the inertia of business-as-usual.

Without a strategic roadmap, cities lose continuity and waste scarce resources on short-term fixes. The transferable lesson is that a clear, forward-looking vision–anchored in foresight and codified in official plans–helps align leaders, agencies and communities on a shared path.

A CITY AT THE CROSSROADS

Digos today stands at a pivotal moment in its history–suspended between promise and peril. It has the ingredients for success: an ambitious vision, a growing population and the recognition of its potential. But without addressing the systemic challenges beneath the surface, it risks becoming another cautionary tale of stalled progress.

The good news is that these challenges are not insurmountable. They are not unique to Digos and they can be overcome through collaborative leadership, meaningful participation and a clear, shared roadmap for the future.

Understanding the state of Digos today is critical because it provides the foundation for imagining its tomorrow.

In the next chapter, we will explore possible futures for Digos City in 2040–scenarios that reveal both the risks of inaction and the transformative power of co-created urban planning.

Part II

THE FUTURES WE WANT

Designing Cities with People, Not Just for Them:
The Power of Collaborative Urban Futures

Archangelo D Alegado, Rec E Eguia and Alan B Alejandrino
ISBN 978-1-83742-369-9
https://doi.org/10.1108/978-1-83742-366-820261007

3

LOOKING AHEAD–SCENARIOS FOR DIGOS IN 2040

"The future does not just happen to us. It is shaped by our choices, our courage and our capacity to imagine something better."

EXPLORING POSSIBLE FUTURES

Every city carries within it not one future, but many. The choices a city makes–or fails to make–in moments of transition ripple forward, shaping lives decades into the future. For a city like Digos, which stands today at the threshold of transformation, understanding these possible futures is not merely an intellectual exercise. It is a moral and practical necessity.

In our research, we worked with community members, local leaders and stakeholders to map out scenarios for Digos City in the year 2040–scenarios that emerged from their own words, fears, hopes and observations. These are not predictions, but plausible pathways, grounded in the city's present conditions, its institutional dynamics and its willingness–or reluctance–to act collaboratively.

Each scenario reveals what is at stake. Together, they invite us to reflect: *What kind of city do we want to become? And what are we prepared to do about it?*

THE RISKS OF BUSINESS-AS-USUAL

Around the world, many cities that fail to confront governance weaknesses drift into stagnation and decline. Political cycles override long-term planning, infrastructure falls behind demand and citizens lose faith in the institutions meant to serve them. The first and perhaps most sobering scenario for Digos reflects this global risk and could be called 'Fragmented Futures: The City Left Behind.'

This is a Digos where nothing fundamentally changes. Political interference continues, infrastructure remains inadequate, marginalised communities remain unheard and the vision of the 'city of choice' becomes little more than a slogan. In this future, roads become more congested and drainage failures increase the city's vulnerability to floods. Public trust in government further erodes, participation dwindles and alienation deepens. Economic growth becomes uneven, concentrated in a few sectors and neighbourhoods, leaving large swaths of the population excluded. Social cohesion weakens and the sense of shared destiny fades. This is the path of least resistance and greatest regret.

This scenario is not unique to Digos. When cities allow politics to override planning, they lock themselves into cycles of underperformance and disillusion. The lesson is clear: the cost of inaction is always greater than the cost of reform.

GRASSROOTS MOMENTUM: POCKETS OF PROGRESS

In many cities, when governments fail to provide leadership, communities step into the gap. Citizens form cooperatives, neighbourhood groups and grassroots coalitions that spark innovation and resilience. These efforts can transform local conditions, but without institutional support they rarely scale. Another possible future for Digos reflects this pattern and could be called 'Grassroots Momentum: Pockets of Progress.'

Here, progress emerges not because of systemic reform but despite its absence. Communities, NGOs and small coalitions create localised projects that improve lives in specific areas–for example, neighbourhood gardens, youth-led initiatives and micro-infrastructure projects. In this future, some neighbourhoods become models of resilience and participation, where citizens prove that change is possible even without strong government backing. However, the absence of coordinated leadership means these successes remain fragmented. Inequality persists as benefits are unevenly distributed and scale is limited. This scenario reflects what happens when citizens care enough to act, but government fails to amplify their efforts.

Grassroots action has value everywhere, but without institutional support it rarely moves beyond isolated projects. The lesson for other cities is clear: governments should not compete with citizen energy but amplify it. They should build scaffolds not silos. Innovation does not trickle down from policy memos; it erupts from the ground. If governments are not creating channels for grassroots brilliance to scale into systemic change, then they are not governing–they are gatekeeping.

EFFICIENT BUT DISCONNECTED: PROGRESS WITHOUT TRUST

Some cities achieve rapid economic and infrastructure gains, yet alienate their people in the process. Investments rise, skylines change and services become more digitised–but when participation is absent, progress feels hollow. This scenario is not unique to Digos; it mirrors the path of many modernised yet disconnected cities.

In this scenario, the city's leadership manages to deliver on infrastructure and economic growth through improved governance systems, partnerships and investments–but without genuinely involving the people, roads are expanded, buildings rise and services become more digitised, and yet citizens feel excluded from the process, with their voices unheard and their needs only partially understood. While the city may look more modern, the social fabric remains frayed. The public sees progress as something done to them, not with them. This is the future of many modernised but alienated cities–efficient, but hollow at the core.

Modernisation without inclusion is a façade. Cities obsessed with efficiency but blind to trust build empires on sand. Progress that excludes is progress that collapses; share it or watch it fracture.

SHARED VISION, SHARED PROSPERITY: THE PREFERRED FUTURE

Cities that succeed in embedding trust and participation into their governance show that prosperity and inclusion can advance together. From Bogotá's participatory budgeting to Helsinki's foresight experiments, the lesson is the same: when people are co-creators of their city's future, growth becomes more resilient and equitable. The most inspiring–and achievable–scenario for Digos carries this same possibility and could be called 'Shared Vision, Shared Prosperity.'

This is the future Digos could realise if it chooses to embrace collaborative foresight, strengthen participation and centre inclusivity in its urban planning. In this scenario, governance becomes transparent and participatory where citizens are treated as co-creators, not spectators. Infrastructure and digital innovations are aligned with the actual needs of the people, creating equitable access to opportunities. Marginalised communities are empowered, not displaced and their voices shape policy. Economic growth becomes inclusive–attracting investments while making guarantees that benefits are distributed fairly. Digos, in this vision, becomes a true city of choice–chosen not only because of its physical amenities, but also because of its spirit of shared ownership, responsibility, trust and possibility.

Collaborative foresight is not just a Digos aspiration–it is a global necessity. The transferable lesson is that inclusive visioning builds legitimacy, and legitimacy is the foundation for lasting prosperity in any city.

Figure 2 illustrates the four plausible scenarios for Digos City in 2040, showing the possible trajectories that emerge from today's choices.

HOW COLLABORATIVE FORESIGHT CAN CHANGE TRAJECTORIES

Across the world, cities are using collaborative foresight to navigate uncertainty and chart resilient pathways. From Helsinki's citizen-led scenario planning to Seoul's long-term visioning processes, the approach shifts urban planning from a technical exercise into a civic contract. Its value lies in

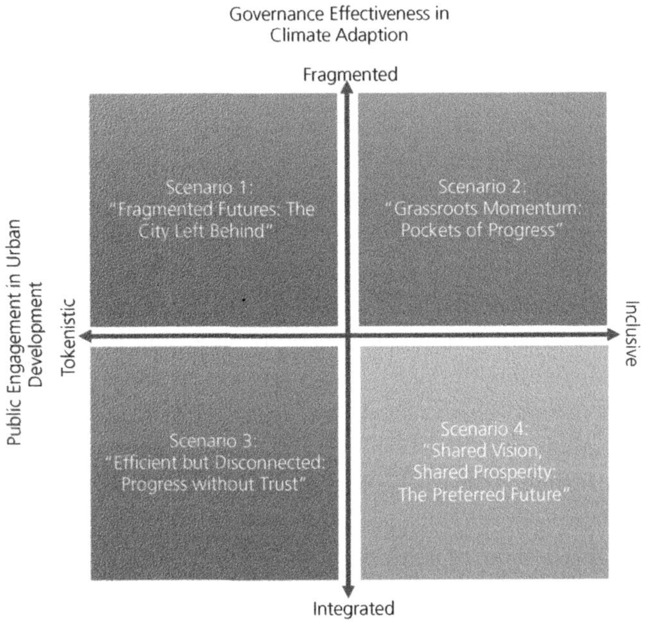

Figure 2 Plausible scenarios of Digos City in 2040

moving the question from *What will happen to us?* to *What future do we want to create together?*

For Digos, this means engaging citizens meaningfully in decisions that affect them, building trust through transparency and accountability, developing a long-term roadmap that aligns infrastructure, policies and programs with the aspirations of its people, and cultivating leadership that listens, facilitates and empowers rather than dictates. The scenarios described in this chapter are not inevitabilities but open trajectories–pathways that can be reshaped if choices are made deliberately and collaboratively today.

The power of collaborative foresight lies in its ability to shift the narrative–from citizens as passive recipients of government action to active participants in shaping their destiny.

This is as true for Digos as it is for any city standing at the threshold of transformation.

A CHOICE TO MAKE

The story of Digos in 2040 has not yet been written. But the choices we make now will determine which story unfolds. Will it be a city left behind, fragmented and disillusioned? Or will it be a city that found the courage and imagination to act together–to design its future with its people, not just for them?

The next chapter of this book begins to answer that question by examining the vision the community itself holds for 2040–the hopes, dreams and values that can guide the way forward.

Designing Cities with People, Not Just for Them: The Power of Collaborative Urban Futures

Archangelo D Alegado, Rec E Eguia and Alan B Alejandrino
ISBN 978-1-83742-369-9
https://doi.org/10.1108/978-1-83742-366-820261009

4

WHAT THE COMMUNITY ENVISIONS

"If you want to know what a city could become, ask the people who call it home."

LISTENING TO THE HEARTBEAT OF THE CITY

The aspirations voiced here–trustworthy services, safe mobility, green space, dignity in development–are echoed in surveys by UN-Habitat and the World Bank in cities across the globe. People consistently ask for fairness, safety, opportunity and inclusion. In Digos, these themes centre on drainage, fair relocation, livelihoods and youth retention.

Urban development plans often speak in the language of infrastructure, budgets and policies, but they rarely capture the heartbeat of a city. To understand what kind of future a city truly needs, you have to listen not just to its leaders, but to its people.

In our research, we spoke with citizens from all walks of life in Digos–city government officials, business owners, community workers and ordinary residents. They spoke not only of what was wrong with the city today, but also of what they dreamed it could become.

Their words revealed more than just opinions. They revealed a deep reservoir of hope, insight and resilience–that is, a vision of a city rooted in dignity, equity and belonging.

WHAT DIGOS RESIDENTS WANT THEIR CITY TO BECOME

When asked to envision their ideal Digos in 2040, residents offered not just aspirations but a heartfelt blueprint for the kind of city they long to belong to.

They described a city where public services are reliable, accessible and free from political interference–a city that responds to the needs of its people with fairness and dignity. They want roads that are not just widened and uncluttered but also safe for everyone: vehicles, pedestrians and cyclists alike. Transportation systems that are organised, efficient and humane, reclaiming streets and sidewalks for people rather than congestion.

Crucially, they insisted that no one should be left behind in the name of development. Informal settlers, relocated families, micro-entrepreneurs, farmers and youth all deserve to share in the city's opportunities. They imagined a Digos where prosperity reaches every neighbourhood, where marginalised sectors are not displaced or silenced but included and supported.

Young people spoke of wanting to stay–of finding meaningful work and building lives in the city rather than feeling compelled to leave in search of better opportunities elsewhere. The city, they said, should invest in industries and skills that make staying not just possible, but desirable.

Green spaces emerged as another powerful aspiration. Residents longed for parks, plazas and public spaces where families could gather, children could play safely and communities could connect. They spoke of wanting to breathe–of a city that values its trees, rivers and open spaces as much as its buildings and streets. In their words, they envisioned a city that balances growth with compassion, technology with humanity and ambition with integrity. They imagined a Digos that honours its history and culture while embracing innovation and progress–a city that is truly resilient and sustainable and that nurtures not just infrastructure but also relationships and identity.

One poignant expression recorded in the dissertation captured it best:

> *"We just want to feel proud of our city–to feel that it is ours and that it cares about us too."*

These aspirations–voiced in focus groups and interviews–reveal not just what residents wish for, but also what they are ready to help build: a Digos they can proudly call their own.

STORIES AND VOICES FROM THE COMMUNITY

Across the key informant interviews (KIIs) and focus group discussions (FGDs), the people of Digos shared not only their frustrations but also their aspirations–stories that reveal both the pain of being excluded and the persistent desire to belong to the city's journey.

> *"We've seen promises before."*

Many residents expressed deep scepticism, shaped by years of broken promises and projects that never delivered what was promised. They spoke of drainage master plans that never left the drawing board, relocation sites that left families feeling uprooted and unsupported, and consultation meetings that felt more like formalities than genuine dialogue.

One resident noted how announcements of plans were made with great fanfare, only to be forgotten months later. This has left many reluctant to trust new initiatives at face value. Yet, even in their disappointment, there remained a quiet hope–a belief that if leaders truly listened, if words turned into action, the city could still change for the better.

"We're ready to help–if they'd just let us."

Participants repeatedly pointed out how decisions made far from their neighbourhoods often failed to account for local realities. Several shared stories of poorly planned projects–drainage works that worsened flooding, or livelihood programs that overlooked the actual needs of the community.

Local leaders, barangay officials and even civil society representatives voiced a similar frustration. "*We know our areas best,*" one key informant said, "*but sometimes it feels like they make decisions in an office without ever stepping foot here.*"

Residents were clear: they did not want to simply watch change happen to them–they wanted to help shape it. They shared examples of grassroots efforts, from community clean-up campaigns to self-organised livelihood projects, demonstrating their willingness and capability to contribute–if only they were invited meaningfully into the process.

"We don't want to be left behind."

Perhaps the most vulnerable voices came from marginalised communities–especially those in informal settlements and relocated families. They shared how modernisation and development often felt like something that happened to them rather than with them. Some feared being displaced again as the city expanded, worried that 'progress' would simply push them further to the margins.

Yet, these same communities also spoke with resilience and vision. They suggested practical initiatives–such as micro-lending schemes, neighbourhood cooperatives and inclusive zoning–that could help them participate in, rather than be victims of, urban growth.

These stories reflect the broader themes of the research: a city where scepticism coexists with hope, where people stand ready to contribute if given the chance and where marginalised voices hold insights that could enrich the city's future, if only those voices are genuinely heard and valued.

The voices of Digos echo a wider truth: communities everywhere want cities that work fairly, safely and sustainably for their people. The transferable lesson is that listening closely to these aspirations is the first step toward designing cities with people, not just for them.

THE POWER OF A SHARED VISION

What became clear throughout the interviews and discussions is that the people of Digos do not lack imagination, nor do they lack commitment. What they lack is a genuine seat at the table–a meaningful role in deciding what their city becomes.

Across conversations, residents consistently expressed that when they are excluded from planning, decisions feel distant and arbitrary. Projects imposed from above–even well-meaning ones–are often met with, as mentioned earlier, scepticism, resistance or apathy because people feel they had no say in shaping them.

Yet, these same discussions also revealed a profound willingness to engage, to contribute and to take responsibility, if given the opportunity. When residents feel that their voices are heard and, more importantly, that their ideas are reflected in the city's policies, budgets and projects, they respond with cooperation rather than resistance. Instead of seeing development as something done to them, they begin to see it as something done with them, and even by them.

This is the power of a shared vision. It does more than inspire; it anchors development in the lived realities of those who experience the city every day. It transforms citizens from spectators into stakeholders–from reluctant recipients of change into its co-authors.

When collaborative planning is practised meaningfully, it cultivates trust where there was once doubt and it builds cooperation where there was once resistance. Citizens take greater care of public spaces when they feel ownership of them. They defend projects and policies they helped design. They become partners not just beneficiaries, bringing with them local knowledge, creativity and resilience that no technical plan alone can deliver.

The research also showed that this shared vision is not simply about agreeing on a list of priorities. It is about creating a sense of belonging–the feeling that "this city is ours and it cares about us too", as one participant put it. This emotional connection is just as vital as the technical details, because it guarantees that plans endure beyond election cycles and adapt to future challenges.

For Digos, the challenge ahead is not to instill imagination in its people–they already have that. The challenge is to build the mechanisms, the culture and the trust necessary to bring that imagination to the centre of its transformation.

WHY IT MATTERS

The vision of Digos as a city of choice will remain hollow unless it is also the *people's choice*, which means

- listening to the quiet voices as well as the loud ones

- treating citizens as partners rather than problems
- ensuring that development does not just happen *to* people but *with* them, and for all of them.

The future of Digos cannot be written in boardrooms alone. It must also be written in barangay halls, markets, classrooms and homes–wherever people imagine a better life for themselves and their neighbours.

In the next chapter, we will examine why public participation so often fails to live up to its promise and how cities can move beyond tokenism to create genuine, meaningful engagement.

Part III

WHY PARTICIPATION MATTERS

Designing Cities with People, Not Just for Them:
The Power of Collaborative Urban Futures

Archangelo D Alegado, Rec E Eguia and Alan B Alejandrino
ISBN 978-1-83742-369-9
https://doi.org/10.1108/978-1-83742-366-820261010

5

BEYOND TOKENISM–MEANINGFUL ENGAGEMENT

"If you're not at the table, you're on the menu."

Participation only works when people's input can change outcomes. Consultation without shared power is theatre–everywhere.

WHY PARTICIPATION OFTEN FAILS

Public participation has become a buzzword in urban planning. Nearly every city today claims to engage its citizens, to listen to their voices and to consult them on decisions. And yet, in practice, participation often fails to live up to its promise. Why?

One reason is that participation is often designed as a performative (or a formality), rather than a genuine conversation. Meetings are held not to hear, but to announce. Consultations take place after decisions have already been made. Surveys are distributed, but their results are never reflected in action.

In many cities, including Digos, participation tends to be

- top-down–driven by what city leaders want to hear rather than what citizens need to say
- exclusive–limited to a small group of stakeholders, leaving others, particularly marginalised communities, out
- short-term–focused on one project at a time, with no effort to build long-term trust or institutional memory.

The consequences of performative participation are visible in the very streets of Digos City. During heavy rains, few urban roads become almost impassable due to flooding–not because of extraordinary weather, but because of ordinary neglect. Despite ongoing road concreting projects, there remains no comprehensive drainage master plan to guide infrastructure development. Roads are paved without proper ground elevation surveys, resulting in uneven surfaces that trap water rather than channel it away. In some areas, newly constructed roads sit higher than adjacent lots, inadvertently redirecting runoff into homes and businesses.

This is not simply a technical oversight–it reflects a deeper governance gap. Projects are implemented in isolation, with little coordination between engineering design, hydrological planning and community input. Residents who experience these problems daily are rarely consulted in ways that influence design. When they are invited to participate, it is often after decisions have been made, or in formats that prioritise visibility over substance.

The flooding issue in Digos is emblematic of a broader pattern: infrastructure built for short-term optics rather than long-term resilience. It illustrates how participation, when reduced to a procedural formality, fails to capture the knowledge and priorities of those most affected.

To address these challenges, urban planning must move beyond symbolic gestures. It must integrate technical rigour with community insight and treat participation as a strategic asset rather than a checkbox.

This pattern is not uniquely local; it is a classic failure mode when engineering proceeds without lived-experience design. The remedy is to bring communities into earlier stages with a visible 'you said, we did' loop.

THE DIFFERENCE BETWEEN INCLUSION AND MANIPULATION

Not all participation is equal and not all of it is genuine.

At its worst, participation becomes a form of manipulation, a way to rubber-stamp decisions and manufacture consent rather than build consensus. As mentioned previously, citizens are invited to meetings only to endorse what has already been decided, with no real power to influence outcomes.

True inclusion, by contrast, is when citizens are given

- information that is transparent and timely
- space to voice concerns, ideas and alternatives
- feedback that shows their input was considered–and explains why some suggestions were or were not adopted
- a sense of shared ownership and responsibility in both successes and failures.

The distinction between inclusion and manipulation can be subtle, but it makes all the difference. Where manipulation breeds distrust, inclusion promotes cooperation. Where manipulation reduces participation over time, inclusion deepens it.

Inclusion also means recognising who has been left out–women, the poor, the youth, persons with disabilities, informal settlers–and making deliberate efforts to bring them into the process. Otherwise, participation risks reinforcing existing inequalities rather than reducing them.

LESSONS FROM GLOBAL CASE STUDIES

Around the world, some cities have moved beyond tokenism and demonstrated what meaningful engagement looks like. While each city has its own context, their experiences offer valuable lessons for places like Digos.

Helsinki, Finland – the functional city

Cities cannot rely on technology alone to solve urban problems. Digital transformation becomes meaningful only when it goes hand in hand with inclusivity–when systems are designed to serve people, not just institutions. Efficiency without equity risks building cities that are technically advanced but socially hollow. The lesson is simple: digital tools expand trust and functionality only when paired with citizen participation and transparent governance.

Helsinki has made this principle explicit by branding itself as 'the most functional city in the world'. The city's Strategy 2017–2021 identified three intertwined pillars: Smart City, Inclusive City and Sustainable City. These were not abstract labels–they structured budget priorities, planning decisions and service design (City of Helsinki, 2017).

Through its Carbon Neutral Helsinki 2035 Action Plan, the city used technology and data to reduce emissions while aligning efforts with residents' daily lives (City of Helsinki,

2018)–for example, transport planning integrated real-time mobility data with citizen feedback, allowing shifts toward low-emission modes without leaving users behind.

Helsinki also embraced participatory budgeting, earmarking €4.4 million annually for residents to propose and vote on projects. This ensured that technology and funding flowed toward needs defined by the people themselves, not just by planners (Björklund and Kronqvist, 2020).

What stands out is how digital systems and participatory structures reinforced each other. Online platforms expanded the reach of engagement, while citizen involvement gave legitimacy to digital projects. As a result, investments in infrastructure, housing and sustainability carried the stamp of both technical quality and democratic accountability.

The lesson is not that Helsinki avoided tension. Challenges persisted–digital literacy gaps, concerns about representation and debates over priorities. But the city addressed these openly, recognising that inclusion was not an add-on but the core condition for legitimacy (Anttiroiko, 2016).

Helsinki's example underscores the risk of digitalisation without genuine participation. Introducing geographic information systems (GIS), early warning systems or paperless governance can streamline services, but unless communities are engaged in shaping and monitoring these tools, efficiency gains may erode trust rather than build it.

A practical step would be to mirror Helsinki's participatory budgeting on a smaller scale–earmarking funds for community-defined projects and requiring that digital platforms be the medium for proposals and updates. This would not only normalise digital participation but also show immediate, visible results of citizen input.

Another application is linking digital infrastructure with climate adaptation. Like Helsinki's Carbon Neutral 2035, Digos could embed participatory monitoring of drainage, flooding and waste management into its systems. Residents would not just report problems but help shape how the city responds, making resilience a shared responsibility.

Digital transformation is sustainable only when coupled with participation. For any city, the test is this: *do residents see technology as a tool they control or a system imposed on them?*

Seoul, South Korea – the social innovation city

Citizen participation cannot depend on ad hoc events or the goodwill of individual leaders. When engagement is treated as a one-off consultation, its effects fade with the political cycle. Participation becomes transformative only when it is institutionalised–embedded in laws, charters and master plans that give citizens predictable influence across administrations.

Seoul is a compelling case because it demonstrates how past mistakes can become the foundation for long-term reform. During the rapid urbanisation of the 1970s and 1980s, Seoul experienced ecological degradation, destruction of cultural heritage and widespread citizen exclusion from decision making (Kim, 2011). Public trust eroded and the city's development model became a symbol of growth without voice.

This legacy pushed Seoul to rethink. In 2013, the city launched the 2030 Seoul Plan, its first citizen-led master plan. Over 100 citizen delegates worked alongside experts and officials to draft long-term priorities. The process was innovative: citizens were not just consulted but given co-decision power in shaping goals such as sustainable transport, inclusive housing and cultural preservation (Seoul Metropolitan Government, 2014).

Two years later, this shift was locked in through the Seoul Urban Planning Charter (2015), a legal framework enshrining ten values, including sustainability, inclusivity and participation. Unlike earlier plans, the Charter was designed to survive mayoral terms–it mandated that citizen participation be a permanent feature of Seoul's planning process (Shin, 2016).

Institutionalisation made the difference. While citizen forums and participatory budgeting mechanisms existed in many cities, Seoul elevated them to charter-level status. This gave residents confidence that their input would not disappear with a change of leadership. It also gave officials a rulebook to justify inclusive practices even against political resistance.

Of course, challenges remain–ensuring inclusivity beyond middle-class groups, managing tensions between expert-driven and citizen-driven priorities and sustaining engagement over time. But Seoul's story shows that when participation is codified, it shifts from symbolic performance to durable governance.

Seoul's trajectory offers a clear lesson for Digos City: pilot projects and consultations are not enough. Unless foresight practices and participatory mechanisms are codified in ordinances or local charters, they will remain vulnerable to political turnover.

A practical step would be to develop a Digos Urban foresight charter, embedding collaborative foresight and participatory planning principles into the city's planning code. This would ensure that engagement is not left to the discretion of individual leaders but becomes a standing requirement across terms.

Another step could be mandating multi-stakeholder planning boards with representation from barangays, youth, women and civil society, whose inputs are formally recorded in decision logs. Like Seoul, the shift would be from 'leaders inviting' to 'citizens guaranteed a seat'.

Participation that depends on goodwill is fragile. Participation embedded in law becomes culture.

Barcelona, Spain – the entrepreneurial smart city

Technological innovation and smart city strategies can generate impressive gains in efficiency, investment and visibility. But without inclusive governance, those same strategies can deepen inequalities and erode social cohesion. The dual lesson is clear: innovation is powerful, but its benefits must be shared. Otherwise, 'smart' cities risk becoming unequal cities.

Barcelona is often hailed as a global pioneer of the smart city movement. In the early 2010s, the city launched hundreds of projects under its 'Barcelona Smart City' brand–integrating sensors, digital platforms and data systems to modernise transport, utilities and public services. Between 2011 and 2015, the city implemented more than 200 projects across 20 programs, generating €85 million for the national GDP and creating over 21 000 jobs (Bakıcı *et al.*, 2013).

Barcelona's digital strategy was structured around three pillars: digital transformation (digitising government operations and services), digital innovation (support for local tech companies and start-ups) and digital empowerment (programs enabling residents to engage with new technologies) (Ajuntament de Barcelona, 2014). These efforts positioned Barcelona as a leader in Europe's network of 'entrepreneurial smart cities', linking urban governance with technology-driven growth (March and Ribera-Fumaz, 2016).

Yet, the city's transformation also carried a warning. In Poblenou, an old industrial district redeveloped under the 22@Barcelona project, global investment and knowledge industries flourished–but many local residents were displaced and benefits were unevenly distributed. Research found that the emphasis on attracting global capital and creative industries often sidelined inclusive participation, leading to socio-spatial segregation (Capel, 2009; Raco and March, 2021).

Barcelona later adjusted its approach. The post-2015 administration shifted the smart city agenda toward digital rights, open data and citizen sovereignty over technology. Policies emphasised not only efficiency but also inclusivity, transparency and resident empowerment (Ajuntament de Barcelona, 2016). This pivot underscored the recognition that innovation must be matched with governance structures ensuring equitable access.

Thus, Barcelona stands as both an inspiration and a caution: innovation can boost prosperity, but without participation at its core, it risks leaving behind the very citizens it is meant to serve.

For Digos, Barcelona's dual story is especially relevant. On the one hand, investments in digital infrastructure–like GIS mapping, early warning systems and online governance–offer a path to modernisation. On the other, if these are pursued purely as technical upgrades without citizen involvement, they risk creating 'two cities': one digitally connected and one excluded.

A balanced step would be to adopt open data platforms that allow residents to access, use and contribute to city information. This could democratise planning knowledge and enable communities to hold government accountable. At the same time, safeguards–such as social impact assessments of major infrastructure projects–would help ensure innovation does not accelerate displacement or inequality.

Barcelona's lesson for Digos is not to reject innovation, but to institutionalise inclusivity as its foundation. Technology can modernise services, but without deliberate attention to equity, progress may come at the cost of social cohesion.

A city becomes truly 'smart' only when innovation and inclusion move forward together.

Singapore – the smart nation

Centralised planning and rapid technological adoption can deliver remarkable efficiency, but efficiency alone is not enough. Technology becomes transformative only when it is aligned with people's needs, embedded in a culture of trust, and applied in ways that preserve inclusivity and social cohesion. The principle is clear: smart governance must be human-centred governance.

Singapore's 'Smart Nation' initiative, launched in 2014, exemplifies this principle. Unlike cities where digitalisation is pursued piecemeal, Singapore adopted a whole-of-nation approach, integrating government, business and citizens under a single strategy (Prime Minister's Office Singapore, 2014).

Three pillars defined the framework (Smart Nation and Digital Government Office, 2018).

- Digital Society – ensuring every resident has access to digital services and opportunities.
- Digital Economy – using technology to attract investment, grow businesses and create jobs.
- Digital Government – embedding technology in governance, from transport to healthcare to housing.

The strategy was reinforced by the Digital Government Blueprint (2018), which committed public agencies to becoming 'digital to the core and serving with heart'. This blueprint emphasised empathy in design, cross-agency data sharing and a strong push for citizen-centric services (Smart Nation and Digital Government Office, 2018).

Singapore also recognised the risks of exclusion. Initiatives such as the Digital Readiness Blueprint (2018) focused on upskilling older residents and vulnerable groups, ensuring no one was left behind (IMDA, 2018). Cybersecurity and data protection frameworks were simultaneously strengthened to build public trust (Smart Nation Singapore, 2014).

The outcome has been a model of highly centralised but inclusive governance: services are efficient, but not at the cost of access or equity. Singapore's success lies in how it framed technology not as an end in itself, but as a tool to enhance quality of life, resilience and trust.

For Digos, the key lesson from Singapore is that technology rollouts must be people first. GIS systems, early warning networks or online permitting platforms should be designed not only for efficiency but also with empathy and accessibility.

Practical steps include building digital inclusion programs for vulnerable groups–especially seniors, rural barangays and low-income households–to prevent a digital divide. Another is institutionalising feedback loops in digital services (e.g. 'report and track' platforms for drainage, waste or safety), ensuring citizens see how their input leads to visible action.

Singapore also demonstrates the value of embedding security and trust from the outset. For Digos, this means developing clear rules for data privacy, accountability in digital governance and transparency dashboards that citizens can access.

A smart city is not measured by how much technology it adopts, but by how effectively that technology expands inclusion, trust and human dignity.

MOVING BEYOND TOKENISM

These examples show that meaningful participation is possible–but it takes deliberate effort to

- design participation as an ongoing process, not a one-time event
- treat citizens as partners, not as boxes to check
- provide feedback that validates their role and sustains their interest
- recognise that participation is as much about building trust as it is about making decisions.

In Digos, moving beyond tokenism means breaking the cycle of superficial engagement and replacing it with authentic, inclusive processes that reflect the city's diversity and aspirations.

Only when citizens feel that their voices matter will they fully invest in building the city they envision. This is as true for Digos as it is for any city. Participation done right does more than inform decisions–it transforms relationships between government and the governed.

In the next chapter, we will examine who holds power in urban development and how cities can shift from controlling change to enabling it, creating an environment where citizens and institutions work together toward a shared future.

Designing Cities with People, Not Just for Them: The Power of Collaborative Urban Futures

Archangelo D Alegado, Rec E Eguia and Alan B Alejandrino

ISBN 978-1-83742-369-9
https://doi.org/10.1108/978-1-83742-366-820261011

6

WHO HOLDS THE POWER?

"Power is not only about who makes the rules, it's also about who gets to speak, who is heard, and who is believed."

POWER DYNAMICS IN URBAN DEVELOPMENT

Urban development is not just about roads, buildings and budgets–it is also about power. Every decision about a city's future reflects a contest of interests, priorities and visions. And in many cities, including Digos, these dynamics often favour those who already hold authority.

In theory, power in a city should be distributed among elected officials, appointed administrators, professionals, private investors and citizens. But, in practice, it tends to concentrate in political elites and powerful institutions.

In Digos, stakeholders consistently expressed that urban development decisions were largely driven by political interests rather than collective needs. Projects are sometimes delayed or redirected because of changes in administration.

Programs benefiting marginalised communities can be sidelined if they do not align with political agendas.

The result? Citizens feel powerless. Many perceive that their voices carry little weight compared with those of politicians or developers. Even barangay officials, who are closest to the people, often find themselves sidelined by city-level decision making.

This imbalance has consequences. When citizens believe their participation is futile, they disengage. When institutions treat public consultation as a performative rather than a genuine dialogue, trust erodes.

In many cities, decision-rights cluster with elites, whether in the form of business coalitions, landholding families or political dynasties. Shifting from control to facilitation requires both rules (such as ordinances that mandate inclusion) and rituals (such as standing citizen forums) that survive election cycles. Without both, power remains concentrated and planning becomes reactive.

OVERCOMING INSTITUTIONAL BARRIERS AND DISTRUST

If cities are to become truly inclusive and resilient, they must confront and dismantle these barriers.

In Digos, stakeholders identified several institutional challenges.

- Centralised decision making. Most decisions are made by a few actors at the top, leaving little room for grassroots input.
- Patronage politics. Development projects sometimes serve as rewards for loyalty rather than responses to real needs.

- Short-term thinking. Planning is often tied to election cycles, with little continuity between administrations.
- Weak accountability. Citizens have limited ways to hold officials and agencies accountable for decisions.

But these barriers are not insurmountable. Around the world, and even within Digos, there are examples of how leadership and institutional reform can rebuild trust. Firstly, by clarifying roles and responsibilities at every level of governance. Secondly, by institutionalising mechanisms for participation that cannot be easily undone by changes in leadership. And thirdly, by investing in transparency, sharing not just what decisions are made but how and why they were made.

Trust grows when citizens can see that their voices matter, when they witness follow through, and when institutions are seen as fair and consistent rather than arbitrary.

This dynamic is hardly unique to Digos. Around the world, cities wrestle with overlapping jurisdictions, weak accountability and the erosion of public trust. The transferable pattern is to clarify institutional roles so citizens know who is responsible; to institutionalise engagement so participation is not discretionary and to publish trade-offs transparently so communities can see why certain decisions were made. These steps reduce suspicion and transform participation into a predictable process rather than a political favour.

HOW CITY LEADERS CAN ENABLE–NOT CONTROL–CHANGE

The most effective city leaders understand that their role is not to dictate the future but to create the conditions for it to emerge collaboratively and inclusively.

This shift in mindset requires moving from control to facilitation. Leaders who enable change focus on creating spaces where diverse stakeholders can contribute meaningfully. They share information openly, even when it is uncomfortable. They recognise the expertise that resides in communities and not just in technical offices. They also avoid using participation as a tool of manipulation, and instead embrace it as a source of better decisions and stronger legitimacy.

In Digos, stakeholders expressed that they want their leaders to be catalysts–people who open doors, remove barriers and empower communities to act.

The underlying message from many communities is clear: they are not asking the city to solve every problem for them. What they ask for is the opportunity to participate meaningfully–to be seen not just as beneficiaries, but as co-creators of urban progress. When leaders transcend the need to control outcomes and instead focus on enabling others, they unlock the creativity, resilience and goodwill of their citizens. This rule of thumb applies everywhere. If people cannot see how their input changed the plan, expect disengagement next time. Leaders who enable rather than control demonstrate that trust is not a soft value but a hard condition for effective governance. In any city, legitimacy grows when decisions are visibly shaped by those most affected.

A SHARED RESPONSIBILITY

Power in a city does not have to be a zero-sum game. When leaders, institutions and citizens share responsibility for shaping the city's future, they create a stronger foundation for development.

The challenge lies in recognising and redistributing power and moving away from an 'us versus them' mentality and toward a shared commitment to the city's well-being.

Digos, like many cities, stands at this threshold. Its future depends on whether its leaders can embrace the humility, courage and vision required to empower their citizens, and whether citizens are ready to step into that power.

In the next chapter, we will explore the critical success factors that make urban transformation possible and how governance and community engagement work together to drive progress.

Part IV

KEYS TO TRANSFORMATION

Designing Cities with People, Not Just for Them: The Power of Collaborative Urban Futures

Archangelo D Alegado, Rec E Eguia and Alan B Alejandrino

ISBN 978-1-83742-369-9
https://doi.org/10.1108/978-1-83742-366-820261012

7

THE CRITICAL SUCCESS FACTORS

> *"Cities are not transformed by chance. They are shaped by choice—and by the courage to act together."*

WHAT MAKES TRANSFORMATION POSSIBLE

Urban transformation is not the result of a single project or charismatic leader, it is the outcome of multiple forces working in harmony. In examining Digos City, it becomes clear that beyond physical infrastructure or economic growth, deeper forces determine whether a city realises its potential.

These forces–leadership, trust, digital infrastructure, innovation, and the alignment of governance and community engagement–are the foundations upon which sustainable change rests. They cannot stand alone. Like strands in a rope, they strengthen each other, weaving resilience into the fabric of the city.

THE ROLES OF LEADERSHIP, TRUST, DIGITAL INFRASTRUCTURE AND INNOVATION

Leaders who facilitate create more capacity than leaders who command.

Leadership emerged as the central catalyst in Digos's journey toward transformation. Effective leaders create an environment where change can take root, not through command and control but through facilitation. Their influence is felt not just in decisions made but in the atmosphere they create–whether inclusive or exclusive, empowering or disempowering.

When leaders act with vision and integrity, they inspire participation, strengthen institutions, and set expectations for fairness and accountability. They are able to balance long-term goals with immediate needs, promoting continuity across political cycles. Conversely, when leadership prioritises personal or political agendas over collective good, the entire system suffers.

In Digos, it became evident that leadership that enables others rather than hoarding authority provides fertile ground for transformation to flourish.

Trust is perhaps the most intangible yet critical element of transformation. In Digos, distrust between citizens and government was a recurring theme–rooted in years of unmet promises, lack of transparency and tokenistic engagement. This erosion of trust made people hesitant to participate and doubtful that their voices mattered.

Rebuilding trust requires a consistent, transparent approach where intentions align with actions. When citizens believe that institutions respect and value their input, they become willing partners rather than skeptical spectators.

Trust also works both ways, not only must citizens trust their leaders but leaders must also trust their citizens enough to genuinely share power and responsibility.

A modern city's ability to thrive increasingly depends on its digital infrastructure, not only for efficiency but also for inclusivity and responsiveness. In Digos, technology has begun to permeate city processes, but the journey toward full digital transformation is still in its early stages.

On paper, the city has embraced digital tools for planning, disaster preparedness, heritage mapping and basic infrastructure upgrades–for example, geographic information systems (GIS) are already used to monitor ecological conditions and map cultural sites. Early warning systems have been implemented to protect vulnerable areas. Partnerships with national agencies, like the Department of Information and Communications Technology (DICT), support efforts to modernise governance.

Yet, these advances remain uneven and institutional readiness is still developing. Based on the Diffusion of Innovations Theory, Digos falls into the Early Majority category of adopters. This means the city has moved beyond scepticism and is deliberately adopting established, proven technologies rather than experimenting with untested innovations. It prefers cautious progression which reduces risk, but this also means it risks falling behind more agile peers if political interference or resistance to change persists.

Several factors explain this measured pace. A significant generational gap within the city's workforce–with employees spanning a wide age range, including some past retirement age–creates challenges in adapting to new systems. Cultural resistance to change and limited internal expertise further slowdown the process.

Still, the groundwork is promising. Initiatives such as the city's partnership with Landbank's institutional internet banking platform for payroll, payments and remittance management reflect a growing willingness to digitalise routine operations. City offices have begun transitioning to paperless systems, implementing online approval processes and strengthening data privacy mechanisms.

To sustain momentum, Digos will need to address its internal barriers through upskilling, leadership that champions innovation, and a cultural shift toward openness and adaptability. Investing in training, securing partnerships and embedding digital literacy across the organisation are essential if the city is to progress beyond being a cautious follower and become a confident leader in digital governance.

Innovation is what allows a city to move beyond the status quo. It represents the willingness to experiment, adapt and create solutions that respond to the community's evolving needs. In Digos, rigid bureaucratic routines often stood in the way of innovation. But there were also signs that, when given space, creative solutions could emerge–from small neighbourhood initiatives to smarter ways of managing resources. Innovation in this context is not just technological. It is also social and institutional–meaning, redesigning how participation happens, rethinking how resources are allocated and imagining new ways of creating value for the community.

It takes courage for a city to break old patterns. But without innovation, even the most ambitious visions risk becoming stale or irrelevant.

Other cities have adopted proven tools first, paired them with upskilling and published performance baselines. However, technology without trust and human capability is just an expensive experiment.

BRIDGING GOVERNANCE AND COMMUNITY ENGAGEMENT

One of the clearest findings from Digos's journey is that the success of its transformation hinges on repairing and strengthening the connection between governance and the community.

The research revealed that development in Digos has historically been dominated by a small circle of decision-makers–often shaped by political interests, administrative priorities and external pressures–while the broader community remained largely on the sidelines. Participation mechanisms existed in name but rarely influenced final decisions. Consultations were often conducted late in the process, limiting the ability of stakeholders to meaningfully shape outcomes.

This pattern created more than policy inefficiencies; it bred a sense of detachment among residents. Many viewed participatory activities as tokenistic, intended to secure compliance rather than invite collaboration. Community members felt that their insights, shaped by direct experience of the city's realities, were undervalued or ignored altogether.

The research underscored that meaningful transformation in Digos requires reversing this dynamic. Community engagement cannot remain peripheral to governance–it must become institutionalised, predictable and inclusive.

Specifically, the findings reveal the following.

- Participation should occur at all stages of planning, not just during approval phases.
- Mechanisms need to ensure that even marginalised and less vocal groups, such as informal settlers, small business owners and youth, are heard.
- Engagement must be structured to survive beyond changes in leadership, embedded in governance systems rather than tied to individual administrators' goodwill.

These findings point to the importance of designing avenues for engagement that are accessible, culturally attuned and aligned with community capacities. Without these, participation risks reinforcing power imbalances instead of dismantling them.

When governance in Digos genuinely reflects the voices and aspirations of its residents–that is, when people see that their contributions shape priorities, budgets and outcomes–a deeper sense of ownership emerges. This sense of shared responsibility strengthens trust, inspires cooperation, and builds resilience against political and institutional disruptions.

Rather than seeing themselves as subjects of development, residents begin to see themselves as co-creators of their city's future, transforming Digos into a true city of choice.

A FRAMEWORK FOR CHANGE

The transformation of a city is a complex process, but it is also deeply human. Leadership provides the direction, trust builds the foundation, digital infrastructure connects people and institutions, and innovation unlocks potential. Together, they enable governance and citizens to work in harmony, forging a city that reflects the aspirations of its people.

Digos's journey shows that these critical success factors are not optional. They are the conditions under which the vision of the 'city of choice' can move from aspiration to reality.

In the next chapter, we will explore lessons from around the world that tell stories of co-creation and innovation that illustrate how other cities have embraced these principles and what they can teach us about building resilient, inclusive and sustainable urban futures.

Designing Cities with People, Not Just for Them:
The Power of Collaborative Urban Futures

Archangelo D Alegado, Rec E Eguia and Alan B Alejandrino
ISBN 978-1-83742-369-9
https://doi.org/10.1108/978-1-83742-366-820261014

8

LESSONS FROM AROUND THE WORLD

"Cities learn not only from within but also by observing what others have dared to try."

STORIES OF CO-CREATION AND INNOVATION

Around the world, cities and communities have demonstrated that genuine co-creation, rooted in inclusivity, empowerment and shared responsibility, can transform not just policies but lives. These stories offer practical insights into how resilience, sustainability and inclusivity are built.

At the heart of these examples lies the idea, echoed by Henri Lefebvre's *'right to the city'*, that real transformation comes when citizens actively inhabit, shape and take responsibility for their urban spaces. This notion of real and active participation shifts residents from being mere recipients of development to rightful co-creators of it.

Principle: capability and ownership drive sustainable tourism (Amadiba experience, Digos application)

Tourism only sustains itself when communities own both the process and the benefits. Without capability and ownership, projects collapse into dependency or exploitation. The presence of natural beauty or cultural heritage alone does not guarantee resilience. What matters is whether people living in those places can meaningfully shape, govern and benefit from how tourism develops (Scheyvens, 1999).

In many contexts, planners have tried to parachute tourism projects into local settings. Too often, residents are treated as labourers or passive 'beneficiaries' rather than as co-creators. The result is predictable: outsiders extract value, while communities are left with degraded environments or weakened social fabric. By contrast, when residents are positioned as active agents, with skills, voice and rights, tourism becomes a source of both pride and livelihood (Scheyvens, 1999).

The Amadiba community in South Africa's Wild Coast illustrates this principle vividly. In the late 1990s and 2000s, Amadiba faced outside proposals for titanium mining and mass-tourism projects that were promoted as economic development but threatened to displace residents, privatise communal land and erode cultural heritage (Mnwana, 2015; SAHRC, 2016).

Community resistance crystallised in the Amadiba Crisis Committee (ACC), which argued that only development rooted in local ownership could safeguard their land and identity (AIDC, n.d.). Rather than accept externally driven projects, residents built their own alternative: Amadiba Adventures, a locally run eco-tourism initiative. The enterprise offered guided hikes, homestays and cultural experiences, with profits channelled through community structures to ensure that benefits were shared broadly instead of captured by elites (Amadiba Eco Adventures, n.d.; Sustaining the Wild Coast, n.d.).

The process required deliberate investment in training. Skills in hospitality, marketing and financial management had to be developed, and tensions sometimes emerged between traditional authorities and newer community organisations. These difficulties underscored that empowerment is not automatic–it has to be cultivated through transparent governance and equitable distribution mechanisms (Mnwana, 2015).

Over time, however, Amadiba Adventures demonstrated that when residents assert control over both narrative and profit, tourism can reinforce identity and resilience. Visitors encountered not a packaged, top-down product but authentic exchanges chosen and curated by the community itself. The outcome was not tourism alone but a stronger sense of ownership, linking livelihoods with stewardship of the landscape (Sustaining the Wild Coast, n.d.).

The Amadiba case also had national significance. In 2018, South Africa's High Court affirmed that communities like Amadiba must give full and informed consent before mining rights can be granted, strengthening the principle of local ownership over land and resources (Centre for Environmental Rights, 2018; ICJ, 2018; Mail & Guardian, 2018). This legal recognition reinforced what the community had long argued in practice: development cannot succeed if imposed against the will of those who live with its consequences.

For Digos City, the Amadiba story highlights both opportunity and risk. With Mount Apo and coastal areas nearby, eco-tourism is often identified as a growth sector. But if it unfolds in ways that sideline local communities–treating them only as service providers or employees–the likely result is resentment and fragile projects that collapse once external funding or leadership shifts.

Instead, Digos can strengthen barangay-level cooperatives and community associations, not just as participants but as decision makers. This requires city ordinances that secure a fixed share of tourism revenues for community reinvestment, plus training programs in enterprise management, digital marketing and environmental stewardship. The city government's role should be that of facilitator, linking communities to technical expertise, seed capital and market platforms, while leaving ownership and governance to the people themselves.

Such a model aligns livelihoods with conservation. When communities see tangible returns from protecting rivers, trails or cultural assets, stewardship follows naturally. This approach also discourages extractive alternatives like mining or unregulated development, offering instead a path where tourism becomes a vehicle for equity and resilience.

Intergenerational co-creation builds civic muscle (Lyon experience, Digos application)

Sustainable urban communities are strengthened when people across generations are actively included–not just as beneficiaries, but as creators. When young people, seniors, families and other groups share space, stories, resources and decision-making, cities become more resilient, socially cohesive and sustainable. This requires design of housing, services and neighbourhoods that purposefully connect diverse people and not buffer them from each other.

In many cities, initiatives that mix age groups–students, retired people, youth, elders and so on–in co-living community facilities, councils or projects show that intergenerational contact reduces isolation, strengthens trust and promotes a sense of belonging. These experiments often succeed when supported by local institutions and policies that ensure continuity, shared responsibility and inclusive design.

Lyon, France, has embedded intergenerational co-creation in several initiatives. A clear example is the One Roof, Two

Generations programme, which pairs older residents with students needing affordable housing. In exchange for low rent, students provide companionship and light support, addressing elder isolation while easing the housing burden for young people (AARP, 2015).

The Lyon Confluence regeneration project further demonstrates institutionalised co-creation. Managed by the public development company SPL Lyon Confluence, it transformed a former industrial area into a vibrant urban district with mixed housing, green public spaces and cultural facilities. Governance was structured to ensure dialogue with residents and long-term coordination across actors, creating spaces where people of all ages could live, work and participate in community life (Darchen, 2019).

Beyond housing and planning, Lyon has also cultivated participatory innovation through the Erasme Living Lab. Operating for more than 20 years, Erasme engages citizens–young and old–in co-designing digital tools, public services and educational projects. It has pioneered methods to include older residents in digital literacy initiatives while testing youth-focused educational technologies, showing how age diversity enriches innovation (Co-VAL, 2019).

At the neighbourhood scale, the 'Le Champ' green infrastructure project illustrates how shared open space supports intergenerational contact. Designed as part of the confluence area, it includes gardens, playgrounds and walkable routes intentionally structured for multiple age groups. Its public realm improvements combined ecological sustainability with opportunities for children, families and seniors to interact daily (AIPH, 2018).

Together, these projects illustrate Lyon's model: when local authorities act as facilitators and create institutional mechanisms, intergenerational co-creation scales up from pilot projects to long-term city policy.

For Digos City, the Lyon experience shows that intergenerational planning is not a luxury but a strategy for resilience. Local housing initiatives could experiment with pairing students or young workers with elderly residents in affordable co-living schemes. Barangay halls, public markets or community centres could be redesigned through workshops where elders, youth and marginalised groups jointly define priorities.

A city-level 'living lab' could also be established to test public service innovations with diverse users–for example, creating youth-friendly and elder-friendly digital tools. The key is to move beyond symbolic consultation and institutionalise mechanisms where different age groups continuously influence design and decision making.

Digos can build civic muscle that is inclusive across generations, reducing fragmentation and strengthening social trust by embedding these practices in ordinances and tying them to the city's development plan.

If we want a city that lasts, let the young speak, let the old guide and let everyone build. That is not idealism–it is durability.

Government as facilitator multiplies social capital (Roskilde experience, Digos application)

Cities thrive when government shifts from being the sole provider of solutions to being a facilitator of community energy. In this role, city authorities create the conditions for citizens, associations and businesses to collaborate, innovate and co-produce services. When government steps back from control and instead enables participation, social capital multiplies and communities become more resilient.

Roskilde Municipality in Denmark has become known for embedding co-creation into local governance. Since the mid-2000s, the city has pursued a strategy of 'from government to governance', focusing on partnerships rather than top-down delivery. Its policies emphasise involving residents, civil society organisations and private actors in shaping projects and services (Sørensen and Torfing, 2016).

A landmark was the Roskilde Festival's collaboration with the city, which demonstrated how a cultural event could serve as a laboratory for co-creation. The municipality worked with festival organisers and volunteers to test ideas around sustainability, mobility and civic engagement. This partnership showed how government can harness citizen enthusiasm and networks, rather than treat them as separate from policy (Sørensen and Torfing, 2016).

Roskilde has also institutionalised co-creation in urban planning. Through its 'city lab' approach, the municipality has engaged residents in shaping neighbourhood renewal, mobility strategies and climate adaptation projects. These labs bring citizens into the design stage, giving them influence over both priorities and solutions. Evaluations show that such processes not only improve project outcomes but also build trust in municipal institutions (Bason, 2018).

The city's 'Roskilde Model of Co-Creation' rests on three pillars.

- Government sets broad goals and frameworks.
- Citizens and stakeholders generate solutions within those frameworks.
- Outcomes are evaluated collaboratively, with feedback loops to adjust strategies (Torfing, 2019).

This approach has been documented as part of Denmark's wider tradition of collaborative governance, but Roskilde stands out for embedding it across sectors–from culture to education, from urban development to welfare services.

Rather than seeing participation as symbolic, the municipality treats co-creation as a method of governing.

For Digos City, Roskilde's lesson is clear: government must position itself as a facilitator rather than a controller. This means setting the frameworks for urban development, such as sustainability goals, zoning rules or resilience targets, and allowing communities, barangays, youth groups and local businesses to propose and shape the concrete solutions–for instance, instead of designing climate adaptation projects purely through technical offices, the city could convene 'community labs' where residents identify flood risks and co-design interventions. Similarly, partnerships with civic groups could turn festivals or cultural events into opportunities for public dialogue and prototyping small-scale innovations, much like Roskilde's use of its festival.

This shift would require changes in institutional culture: city officials trained to facilitate workshops rather than deliver directives, ordinances that institutionalise co-creation as a governance method, and transparent monitoring systems that keep all actors accountable. The payoff would be a stronger reservoir of social capital and a broader base of ownership for Digos's development trajectory.

When government facilitates rather than controls, social capital multiplies and cities gain both legitimacy and resilience.

From these diverse contexts, a common principle emerges: communities that are empowered and supported make meaningful contributions to shaping their cities' futures. Our research distilled several qualities that underpin these successes, namely

- authentic engagement and trust-building–beginning with honest dialogue, listening and transparent communication
- capacity building and education–equipping citizens with the knowledge, skills and confidence to participate meaningfully

- inclusive stakeholder involvement–ensuring representation from marginalised, less vocal or vulnerable groups
- shared responsibility and ownership–promoting solidarity among citizens, businesses and officials
- innovative methods–using participatory design workshops, co-design charrettes and digital tools to amplify creativity and accessibility.

THEORIES ILLUMINATING THIS RESEARCH AND THOSE SUCCESSFUL STORIES

This research and successful stories above illustrate and validate several key theoretical frameworks, which help explain why certain approaches succeed in promoting inclusion, trust and resilience. These are as follows.

- Lefebvre's Right to the City–Henri Lefebvre argued that urban dwellers have not just the right to inhabit the city but also to participate in shaping its policies, spaces and identity. This goes beyond access to resources; it is about claiming agency in the city's transformation (Bailey, 2020; Purcell, 2014).
- Amartya Sen's Capabilities Approach–Sen reminds us that development is meaningful only if it expands people's capabilities; their freedoms to make choices, take action and pursue lives they have reason to value. Empowerment is thus measured not just by material gain but by enhanced agency (Clark, 2005).
- Social Capital Theory–This framework highlights how trust, shared norms and networks within a community enable collective problem-solving and build resilience.

Cities that promote social capital find it easier to mobilise citizens in times of crisis and cooperation (Bolton, 2005).

- Co-Creation Theory (Lund, OECD and others)– Co-creation frames citizens and governments as collaborators rather than separate actors. Through shared responsibility, innovation and reciprocal power, both parties create solutions that are more inclusive and effective (Lund, 2018).
- Arnstein's Ladder of Participation–Sherry Arnstein's classic model illustrates the spectrum of citizen involvement, from manipulation at the bottom to full citizen control at the top. Too often, participation stops at tokenism (informing or consulting) without real power-sharing. The cases demonstrate the transformative potential of climbing higher on this ladder toward partnership and empowerment (Arnstein, 1969).
- Habermas' Theory of Communicative Action–Jurgen Habermas emphasises the importance of rational, inclusive dialogue where participants engage as equals to reach understanding and consensus. Development that emerges from such deliberation, rather than imposed mandates, is more likely to be trusted, legitimate and sustainable (Habermas, 1984).

These theories shed light on why collaborative approaches, rooted in dialogue, trust, shared power and enhanced capabilities, create more resilient, inclusive and vibrant cities.

A SHARED PATH FORWARD

Amadiba, Lyon and Roskilde demonstrate that even in vastly different contexts–rural, urban, developed or developing–the principles of empowerment, inclusion and creativity can create remarkable outcomes.

These stories also serve as cautionary tales: co-creation is not automatic, nor is it easy. It requires deliberate design, sustained commitment and the humility to let go of top-down control. But when done right, it produces more than just better policies–it builds stronger communities, more resilient cities and a deeper sense of shared destiny.

For Digos and cities like it, these lessons affirm that the road to becoming a city of choice does not lie solely in infrastructure or policies but in the people's ability to shape their own future.

In the next chapter, we will bring together all these insights to outline a practical, actionable framework that cities can use to design more inclusive, resilient and sustainable urban futures.

Part V

A FRAMEWORK FOR ACTION

Designing Cities with People, Not Just for Them:
The Power of Collaborative Urban Futures

Archangelo D Alegado, Rec E Eguia and Alan B Alejandrino
ISBN 978-1-83742-369-9
https://doi.org/10.1108/978-1-83742-366-820261015

9

BUILDING THE STRATEGIC FRAMEWORK

"Cities don't become inclusive, resilient and sustainable by accident–they are led there."

Cities are not transformed by isolated projects or fleeting leadership terms. They are transformed when there is a framework that holds together vision, participation and sustained action. Projects may deliver quick wins, but frameworks provide continuity, clarity and resilience–especially when political cycles, economic pressures and social expectations shift.

In this chapter, two elements come together. The Collaborative Foresight Model provides the 'how'–a process through which citizens, leaders and institutions imagine and plan their futures together. The Five Strategic Pillars provide the 'what'–the foundational conditions that every city must strengthen to turn foresight into real transformation. Together, they form a roadmap that is both practical and transferable, rooted in Digos but applicable to any city striving for inclusive and sustainable change.

THE COLLABORATIVE FORESIGHT MODEL EXPLAINED

The Collaborative Foresight Framework provides Digos City with a deliberate, actionable roadmap for realising its 2040 vision of becoming the city of choice. The framework is rooted in both theory and empirical findings, designed to tackle the city's unique challenges while building on its strengths.

The model integrates

- Political, Economic, Social, Technological, Environmental, Legal (PESTEL)-based environmental scanning (to understand context)
- futures thinking tools (like scenario analysis and causal layered analysis)
- transformation process guided by two critical success factors
    - governance and community engagement
    - innovation and economic development.

These success factors give rise to Five Strategic Pillars, each requiring not only technical interventions but also profound shifts in leadership mindset and practice. The pillars are not independent silos–they reinforce each other–and the role of leadership is to weave them into a coherent, lived strategy.

THE FIVE STRATEGIC PILLARS

Pillar 1: Catalytic Leadership

Leadership is the cornerstone of urban transformation. But as the research revealed, leadership in Digos has often been top-down, transactional and reactive, which reinforce distrust and exclusion.

The Catalytic Leadership pillar calls for a different kind of leadership. One that enables others to act rather than hoarding authority. One that builds trust and continuity beyond political terms.

To articulate this pillar, Digos City's leadership must do the following.

- Shift from directing to facilitating by creating spaces for dialogue, deliberation and negotiation, where citizens feel their voices matter. This aligns with Habermas' Communicative Action, which insists decisions should emerge from rational, inclusive dialogue.
- Demonstrate continuity of vision and resist the temptation to discard predecessors' plans for political gain. The city's strategic direction should transcend electoral cycles.
- Model integrity and transparency. Leadership must not only speak about inclusion but demonstrate it through action–involving communities in decisions and clearly explaining trade-offs.
- Empower mid-level and barangay leaders. Decentralise not just authority but also resources, enabling grassroots leaders to respond to local needs effectively.

Leaders must become catalysts–that is, sparking energy, aligning stakeholders and making sure that development is not about their legacy but about the city's collective future.

Pillar 2: Co-created Communities

For too long in Digos, participation has been perfunctory–consultative meetings held after decisions were already made, with outcomes predetermined. This pillar insists that communities must not just be informed but actively co-create policies and projects.

Here, leadership must articulate genuine partnership with residents in the following ways.

- Institutionalise participatory planning by embedding participation mechanisms in formal governance processes, guaranteeing that decisions at every level invite input.
- Move up Arnstein's Ladder of Participation by moving beyond consultation, toward partnership or even citizen control, where feasible. The goal is not just to listen but to share decision-making power.
- Include marginalised groups. Leadership must guarantee that informal settlers, youth, women, persons with disabilities and micro-entrepreneurs have seats at the table. This requires active outreach and culturally sensitive engagement.
- Support urban living labs by facilitating spaces where citizens and city staff experiment with solutions together–from waste management to public space design.

Leaders must recognise that co-created communities build social capital, trust and resilience–outcomes no amount of technical expertise can substitute.

Pillar 3: Future-proof Economy

Economic development in Digos has historically concentrated on a few sectors and benefitted certain groups more than others. This pillar aims to diversify the economy, making it inclusive and resilient.

Leadership's role here is to articulate a vision of the economy that achieves the following.

- Balances investment attraction with equity, attracting large investors while also supporting small enterprises and informal workers.
- Enables capability-building by providing training, credit access and mentorship to entrepreneurs–reflecting Amartya Sen's capabilities approach to development.
- Prepares for disruption by building sectors that can withstand shocks, such as renewable energy, 'regenerative' economies and digital services while investing in resilient supply chains.
- Invests in connectivity by improving transport and digital infrastructure to enhance access to markets and services for all residents.

Leaders should champion a narrative that economic growth must serve the city's people, not the other way around and commit to policies that support broad-based opportunity.

Pillar 4: Eco-friendly Urban Design

The city's rapid growth has strained its ecological resources, diminished public spaces and endangered cultural heritage. This pillar demands that development respects both environment and identity.

Leaders should articulate and enforce a vision that achieves the following.

- Designs for compact, walkable and green neighbourhoods, reducing congestion and improving quality of life.
- Prioritises climate resilience by integrating green infrastructure (parks, rain gardens, flood plains) and renewable energy into urban planning.
- Protects heritage by identifying and preserving historic districts and cultural landmarks that strengthen the city's sense of identity.
- Institutionalises ecological consciousness by embedding sustainability standards in all city projects and educating residents about their role in stewardship.

Leadership here must communicate that a city's greatness is measured not only by its buildings but also by the well-being of its environment and the stories of its people.

Pillar 5: Agile Policy Ecosystem

Digos City's governance has historically been rigid and slow to respond to evolving challenges, which is a symptom of bureaucratic inertia and short-term political priorities.

The Agile Policy Ecosystem pillar calls for leadership to promote institutional adaptability and foresight. To do this, leaders should address the following.

- Embed foresight and feedback loops into planning by institutionalising regular scenario-planning exercises, policy reviews and citizen satisfaction assessments.
- Utilise data-driven decision making and invest in digital governance systems to track performance and adjust in real time.

- Align local with national to navigate the tension between national mandates and local realities, advocating for autonomy when necessary.
- Protect institutional memory, ensuring that knowledge and expertise are retained across political transitions through documentation and professional civil service.

Leaders must champion a culture where policies are not static documents but living tools–adjusted as circumstances change.

Notably, agility is meaningless without feedback. Cities need to measure whether policies are working and adapt when they are not. Globally, urban resilience frameworks emphasise clear metrics to track progress, compare across contexts and sustain accountability (OECD, 2020; UN-Habitat, 2020).

Key domains and indicators include

- governance and participation or the percentage of policies with citizen input and frequency of published decision logs
- service delivery or the turnaround time for permits, grievance resolution rates, percentage of services digitised
- equity and inclusion or the share of budget directed to vulnerable groups and gender-disaggregated access to services
- sustainability or the percentage of waste recycled, green space per capita, carbon emissions per household
- resilience or the disaster response time, percentage of population covered by early warning systems.

For cities like Digos, starting small is enough. Choose 5–7 indicators, publish them on a public dashboard and update annually. What matters is not volume but credibility.

In other words, people must see that data drives decisions, not rhetoric.

WHY LEADERSHIP MUST OWN THIS FRAMEWORK

The Collaborative Foresight Framework will not implement itself. Leadership must actively embody and communicate each pillar, making it clear that transformation is not just about infrastructure, but also about relationships, culture and courage.

City leaders must learn to speak and act differently, moving beyond control to facilitation, beyond rhetoric to inclusion, beyond the present to the possible. They must guarantee that every department, every barangay and every citizen understand their role in this shared journey.

This is leadership as stewardship of trust, of vision and of the collective future.

The strategic framework (Table 1) summarises the vision, critical success factors and strategic pillars (Figure 3) developed through this research, providing a clear roadmap for leaders and communities to collaborate toward the preferred future.

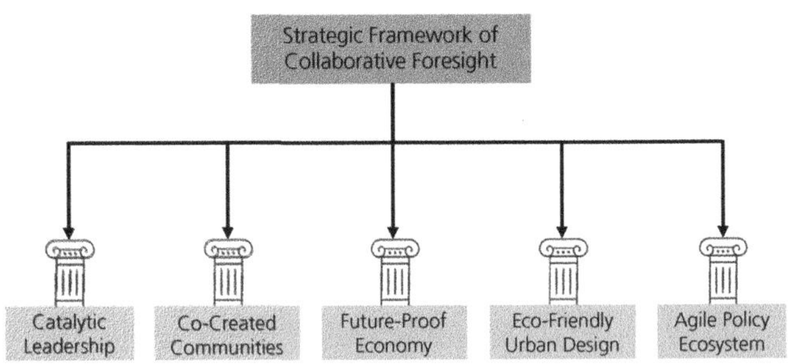

Figure 3 Five pillars of Digos City Transformation in 2040

Table 1 Actionable recommendations anchored on five strategic pillars

No.	Pillars	Policy directions	Program agenda
1	Catalytic Leadership	Strengthen transparent and accountable governance where leaders do not just govern from the top down but co-create solutions with the people.	• Forge the next generation of power players. • Take governance digital. • Zero tolerance on corruption and full commitment to results.
2	Co-Created Communities	Institutionalise participatory urban development where decision making is not locked away in government offices but happens on the streets, in community hubs and through real-time digital platforms.	• Set up an urban living lab where tech minds, community leaders and everyday citizens collaborate to design neighbourhoods. • Make cultural and social programs impossible to ignore. • Bet big on grassroots power by launching a rapid-funding initiative. • Go all-in on 'macro-surgery' for full-scale urban transformation.
3	Future-Proof Economy	Ditch the one-track economy.	• Turn Digos into the next innovation hotspot. • Rethink transport because gridlock and outdated mobility are deal-breakers. • Make heritage and tourism an unstoppable economic engine.

Table 1 (Continued)

No.	Pillars	Policy directions	Program agenda
4	Eco-Friendly Urban Design	Promote sustainable city planning through eco-innovation zones and climate-responsive urban design.	• Design and implement carbon-offset initiatives and green infrastructure projects. • Strengthen flood mitigation strategies through nature-based solutions and resilient urban design. • Introduce waste-to-energy programs and expand regenerative economy initiatives.
5	Agile Policy Ecosystem	Ditch outdated, slow-moving governance.	• Deploy real-time urban monitoring systems for evidence-based policymaking. • Cut through the red tape and put innovation in the driver's seat. • Implement open data governance systems that put real-time public spending, project updates and city performance metrics at everyone's fingertips.

Having traced Digos City's journey and outlined the principles of collaborative urban futures, we now turn to the practical tools that planners and practitioners can use to bring these principles to life–tools that help cities see clearly, think ahead and act together.

Designing Cities with People, Not Just for Them: The Power of Collaborative Urban Futures

Archangelo D Alegado, Rec E Eguia and Alan B Alejandrino

ISBN 978-1-83742-369-9
https://doi.org/10.1108/978-1-83742-366-820261016

10

TOOLS FOR PLANNERS AND PRACTITIONERS

"A city's ability to change begins with its ability to imagine. And imagining well is a skill–one that can be learned, practiced and shared."

The Collaborative Foresight Framework equips city planners, engineers, administrators and community leaders not only with a strategic vision but also with a practical toolkit for realising it.

These tools–used in our research and tested through community engagement–help illuminate challenges, explore futures and develop actionable strategies. Importantly, they are not 'one size fits all'; they must be tailored to the specific realities of a city like Digos.

This chapter provides planners and practitioners with a clear explanation of each tool and guidance for applying them effectively in local contexts. But before any technical tool is applied, there is one essential first step: building futures literacy among those who will use them.

It is important to remember that these tools are not meant to be applied all at once, nor are they rigidly sequential.

Each tool–whether futures literacy, PESTEL analysis, scenario planning, causal layered analysis or factor analysis–can stand alone to address specific needs, or they can complement one another when a deeper, multi-layered understanding is desired. Leaders and planners should feel empowered to choose which tools are most appropriate for their context, resources and stage of planning. What matters is not the number of tools used, but the thoughtfulness with which they are applied to help the city see clearly, think creatively and act decisively toward its preferred future.

FUTURES LITERACY: BUILDING THE CAPACITY TO ANTICIPATE

Futures literacy, as developed by UNESCO and expanded by Riel Miller (2007) in his seminal work 'Futures literacy: a hybrid strategic scenario method', refers to the ability of people to understand and use the future–not to predict it, but to explore multiple possibilities, question assumptions and imagine preferred outcomes.

Futures literacy workshops serve as a foundation for all subsequent tools, allowing stakeholders to loosen rigid, short-term thinking and open themselves to alternative futures.

PESTEL ANALYSIS: UNDERSTANDING CONTEXT

PESTEL analysis helps planners systematically scan their government's political, economic, social, technological, ecological and legal environment.

In our research, this analysis revealed how Digos was shaped by national political interference, dependence on national funds, demographic pressures, weak digital infrastructure, environmental degradation and outdated legal frameworks–all of which informed the design of the Strategic Framework.

How local government leadership can use it

Local governments vary widely, some have strong technical teams and strong budgets, others are stretched thin and rely heavily on national government support. No matter where your municipality falls on this spectrum, PESTEL analysis can be contextualised to uncover actionable insights about an environment.

Here is a step-by-step guide tailored for diverse settings.

Step 1: start with a collaborative workshop

1.1 Convene a PESTEL workshop at the beginning of each planning cycle.

1.2 Invite a diverse group, such as department heads, barangay leaders, civil society organisations, business representatives, youth and marginalised groups.

1.3 In resource-constrained settings, keep the group smaller but make sure it is representative of key sectors and communities.

Step 2: update regularly as context changes

2.1 Recognise that external forces, such as climate policies, technological advances, migration patterns or legal reforms, are dynamic.

2.2 Schedule an annual review of the PESTEL findings to incorporate new developments.

2.3 In fast-changing environments or crisis situations, conduct more frequent 'mini-PESTELs' focusing only on the most relevant dimensions.

Step 3: use PESTEL beyond diagnosis

3.1 Do not treat PESTEL as just an academic exercise; use it to identify leverage points where leadership can act.

- Example: Emerging technologies? Invest early in digital literacy.
- Example: Shifting legal mandates? Advocate for local needs within national frameworks. Scholars call it 'legal contextualisation'.

3.2 Focus on opportunities as much as on uncertainties (or threats) and think creatively about how to turn challenges into drivers of innovation.

Guidelines for all contexts

(*a*) Ground the analysis in local realities by going beyond national data by collecting insights from local records, direct observations and lived experiences.

(*b*) Simplify when needed. If time or capacity is limited, prioritise the dimensions most relevant to your city's immediate concerns.

(*c*) Facilitate participation effectively by using tools like sticky notes, breakout groups or online surveys if in-person participation is not possible, to make sure everyone's voice is heard.

(*d*) Document and share by summarising findings in a simple, visual format that can be shared widely–so everyone understands the city's environment and priorities.

SCENARIO PLANNING: EXPLORING POSSIBLE FUTURES

Scenario Planning helps stakeholders imagine and articulate multiple plausible futures–including risks and opportunities–rather than relying on a single forecast.

In Digos, our research presented four scenarios for 2040.

- Fragmented Futures: The City Left Behind.
- Grassroots Momentum: Pockets of Progress.
- Efficient but Disconnected: Progress Without Trust.
- Shared Vision, Shared Prosperity: the Preferred Future.

Each scenario emerged from literature review, desk review, key informant interviews and focus group discussions, which illustrates how choices and policies today shape the future.

Step 1: define the purpose and scope

1.1 Clearly articulate why you are conducting scenario building: is it to test your city's strategic plan? to prepare for crises? to guide investments?

1.2 Decide the time horizon–in Digos, 2040 was chosen to push thinking beyond short political terms.

1.3 Select the focus areas, such as urban mobility, housing, climate resilience or inclusive governance so scenarios are grounded in real concerns.

Step 2: assemble a diverse group

2.1 Invite representatives from across sectors: government departments, barangays, business community, civil society, faith groups, youth, marginalised residents.

2.2 If resources are limited, aim for diversity even in a smaller group so that voices of the vulnerable are included.

Step 3: gather key inputs

3.1 Conduct a PESTEL analysis to map the forces shaping your city.

3.2 Identify the critical uncertainties or factors you cannot predict but that would profoundly affect the future (e.g. national political shifts, technological breakthroughs, natural disasters).

3.3 Agree on the certainties or trends that are almost sure to continue (e.g. population growth, climate change impacts).

Step 4: build the scenarios

4.1 Choose two or three axes of uncertainty that matter most, such as governance quality (strong as against weak) and public trust (high as against low).

4.2 Use these to create 3–4 contrasting scenarios that explore a range of plausible futures, from highly desirable to highly undesirable.

4.3 Flesh out each scenario with narratives.

- What does daily life look like?
- What are the biggest challenges?
- What opportunities emerge?
- How do governance, economy, environment and society interact?

Step 5: use scenarios to promote dialogue

5.1 Present scenarios visually and narratively to stakeholders use storyboards, roleplay or simple infographics.

5.2 Facilitate discussions about

- Which scenario seems most likely today?
- Which is most desirable?
- What actions could shift us toward the preferred future?
- What risks must we prepare for if less desirable futures materialise?

Step 6: develop strategic responses

6.1 Use scenarios to stress-test current plans: would this policy still work, for example, in Scenario 3?

6.2 Identify no-regret actions or interventions that make sense in all futures.

6.3 Craft contingency plans for less desirable but plausible futures, strengthening resilience.

Step 7: institutionalise scenario planning

7.1 Schedule regular scenario exercises, say every 3–5 years to keep plans adaptive and relevant.

7.2 Train staff to facilitate scenario workshops internally.

7.3 Integrate findings into annual development plans, budget priorities and communication strategies.

Guidelines for all contexts

(*a*) Start simply but meaningfully. Do not get bogged down in jargon; focus on telling clear stories about plausible futures.

(*b*) Make it tangible. Use local examples, maps and even drawings to bring scenarios to life.

(*c*) Encourage honesty by creating a safe space where participants can speak openly about their fears and frustrations without judgement.

(*d*) Document everything. Record key insights, photographs of workshop outputs and participant feedback for transparency and institutional memory.

(*e*) Follow up after the workshop, share a concise, illustrated summary with all participants and the wider community to maintain momentum and trust.

Why scenario planning matters for leadership

By leading scenario-building exercises, leaders signal a willingness to confront uncertainty, listen to their communities and plan beyond electoral cycles. They show that the future is a shared responsibility and that leadership means enabling others to see their role in shaping it.

CAUSAL LAYERED ANALYSIS (CLA): DIGGING DEEPER

Causal layered analysis (CLA) (Inayatullah, 2004) helps leaders and stakeholders go beyond surface-level problems to uncover and address the deeper systemic, cultural and even subconscious drivers that keep these problems in place.

Instead of asking only *what is happening*, CLA asks *why is it happening, what sustains it and what beliefs or stories allow it to persist.*

In Digos, CLA revealed that certain 'rational' policies, like prioritising car-oriented roads, were underpinned by deep-seated cultural beliefs and myths, such as equating wider roads with progress and prosperity. Without addressing these underlying layers, technical solutions often failed or met resistance.

Here is how leadership can systematically use CLA in their context.

Step 1: select the issue

1.1 Identify a persistent, complex issue that has proven difficult to resolve despite repeated efforts. Examples can include traffic congestion, resistance to heritage preservation or citizen disengagement.

1.2 Frame it neutrally so stakeholders feel safe to discuss it openly.

Step 2: assemble a representative group

2.1 Include those directly affected (residents, business owners, commuters) and those responsible for policy (department heads, planners, engineers).

2.2 Ensure diverse perspectives: women, youth, informal workers and marginalised communities often see aspects that others overlook.

Step 3: explore the four layers

Facilitators guide participants through the four CLA layers, documenting insights at each stage.

Layer 1: litany (what everyone sees)

- What are the visible symptoms?
- Example: "*The city is always flooded.*"
- Record facts, statistics, headlines.

Layer 2: systemic causes

- What institutional, structural or policy factors contribute?
- Example: poor drainage maintenance, unregulated construction, weak enforcement of zoning laws.

Layer 3: worldview/discourse

- What assumptions or cultural norms underpin these systemic factors?
- Example: belief that waterways are acceptable dumping grounds; perception that development always means construction.

Layer 4: myth/metaphor

- What deep stories, symbols or metaphors sustain these beliefs?
- Example: "*Taming nature shows progress*", or "*A clean-up only matters if visitors see it*".

Step 4: synthesise and reflect

4.1 Map all layers in one visual so participants can see connections.

4.2 Discuss: Which layers have we been addressing so far? Where do we need to focus more?

4.3 Highlight how addressing deeper layers can unlock more sustainable solutions.

Step 5: develop interventions across layers

5.1 Litany: improve reporting, communication and monitoring of the issue.

5.2 Systemic: reform policies, strengthen enforcement, redesign programs.

5.3 Worldview: launch awareness campaigns, community dialogues, education programs to shift assumptions.

5.4 Myth/metaphor: create new narratives through art, stories and public symbols that reframe the city's identity and aspirations.

Guidelines for all contexts

(*a*) Use real language and avoid jargon when engaging citizens. Instead of "*worldview*", you can say "*what we assume is true*" or "*how we see the world*".

(*b*) Create a safe space. Deeper layers often surface sensitive beliefs or uncomfortable truths; therefore, facilitators must promote trust.

(*c*) Be patient. Change at the deeper levels takes time and persistence.

(*d*) Link findings to strategy. Do not let insights sit on paper. Instead, incorporate them into urban plans, communication strategies and project designs.

(*e*) Celebrate progress by sharing stories where shifting narratives led to real change.

Why CLA matters for leadership

For leaders, CLA is a tool to see the whole picture, not just what is measurable but also what is meaningful. It equips them to

- recognise resistance as a signal of deeper dynamics rather than mere obstruction
- design interventions that resonate with people's values and identity
- assist in aligning policy and culture, which increases the likelihood of success.

FACTOR ANALYSIS: MAKING SENSE OF COMPLEX DATA

Factor analysis is a statistical method that helps make sense of complex, overlapping data by uncovering underlying dimensions, called factors, that explain patterns in responses.

In our research, factor analysis was not simply used to analyse generic attitudes but to uncover the critical success factors that drive urban transformation in Digos. Specifically, it distilled 552 survey responses about governance, participation, trust, innovation and development priorities into two central factors.

(*a*) Governance and community engagement.

(*b*) Innovation and economic development.

These two factors became the foundation for the five strategic pillars of the Collaborative Foresight Framework.

So, for leadership in similar contexts, factor analysis is not just a technical exercise–it is a way to identify what truly matters to people and what levers of change have the most potential impact.

Step-by-step guide

Step 1: define the objective

1.1 Clarify what you want to learn. The objective should be to determine which underlying factors explain stakeholders' attitudes and priorities about a particular subject (e.g. urban transformation).

1.2 Frame questions accordingly by focusing on areas relevant to governance, trust, participation, innovation and development goals.

Step 2: design the survey instrument

2.1 Base the questions on themes surfaced in earlier workshops (e.g. futures literacy, PESTEL, CLA).

2.2 Keep the language clear and locally appropriate to ascertain comprehension across diverse groups.

2.3 Use Likert-scale questions (e.g. strongly agree → strongly disagree) for attitudes and priorities.

2.4 Make sure the survey covers a broad enough scope but avoids redundancy.

Step 3: collect data from a representative sample

3.1 Aim for broad coverage, including local government employees, barangay leaders, business owners, citizens from various socio-economic and age groups.

3.2 Use quota sampling technique to make sure there is adequate and balanced representation.

Step 4: analyse the data using factor analysis

4.1 Work with a statistician or local university to run the factor analysis using software like SPSS.

4.2 Look for factors that explain the majority of variance and make conceptual sense.

- The factor analysis should result to at least two coherent factors, each grouping together several related indicators–for example, in our research governance and community engagement indicators were trust, transparency, inclusiveness and participation, while innovation and economic development a readiness for change, willingness to adopt technology and entrepreneurial outlook were some of the indicators.

Step 5: interpret and use the findings

5.1 Present the findings visually and show which variables cluster under which factor.

5.2 Use the factors to prioritise strategic actions and allocate resources where they will have the greatest effect.

Guidelines for all contexts

(*a*) Involve stakeholders, even if analysis is technical. Share findings in plain language with the community and validate them through feedback.

(*b*) Combine with qualitative insights. Use what you have learned from CLA and scenario planning to make sense of the factors.

(*c*) Ensure diversity in data collection. Do not rely solely on senior staff or politically connected groups.

(*d*) Focus on actionability. Do not stop at reporting findings; tie them directly to planning priorities and budgets.

(*e*) Repeat periodically, say every few years, to track progress and adjust priorities.

Why factor analysis matters for leadership

Leaders often face an overwhelming flood of opinions, complaints and requests. Factor analysis enables them to cut through the noise and identify the essential drivers of transformation in their specific context–those factors that, if strengthened, will have the widest and most meaningful impact.

In Digos, this method revealed that building trust and community engagement was just as critical as promoting innovation and economic development–enabling leadership to design a balanced, targeted strategy rather than a scattershot one.

Designing Cities with People, Not Just for Them: The Power of Collaborative Urban Futures

Archangelo D Alegado, Rec E Eguia and Alan B Alejandrino
ISBN 978-1-83742-369-9
https://doi.org/10.1108/978-1-83742-366-820261021

TOOLKIT FOR COLLABORATIVE URBAN FUTURES–AT A GLANCE

This toolkit (Table 2) brings together five core methods of the Collaborative Urban Futures approach, drawn directly from the research. Each tool is explained briefly, with guidance on its purpose, how it was used in the study, how leaders can adapt it locally and practical tips for implementation. These tools can be used independently or combined for greater depth, depending on the needs and capacity of a local government. The table below serves as a quick reference for planners, leaders and communities committed to designing their future collaboratively.

In the closing chapter, we step back to reflect on the journey of Digos, and issue a call to action for cities everywhere: to design with their people, not just for them.

Table 2 Toolkit for collaborative urban futures

Tool	Purpose	How it was used in the study	How leadership can apply it locally	Practical tips
Futures literacy	Build the capacity to imagine and explore alternative futures.	Helped stakeholders uncover assumptions and envision Digos in 2040 beyond short-term concerns.	Convene diverse groups to reflect on hopes, fears and assumptions; use storytelling and creative exercises to surface insights.	Keep it simple, avoid jargon and focus on real stories to make futures thinking tangible.
PESTEL analysis	Analyse the political, economic, social, technological, environmental and legal context.	Identified external forces and challenges shaping Digos' development.	It can be done via desk review. You may also organise regular workshops with representatives from government, business and community to map trends and uncertainties.	Focus on local realities, not just national data. Use visuals and clear language to engage everyone.
Scenario planning	Explore and prepare for multiple plausible futures; test strategies.	Created four scenarios for Digos in 2040, from undesirable to preferred.	Facilitate workshops to build and discuss scenarios, then stress-test plans against them.	Make scenarios vivid and relatable with names, stories and visuals. Focus on actionable steps.
Causal Layered Analysis (CLA)	Uncover deep cultural, systemic and mythic causes of surface problems.	Revealed that resistance to change was rooted in cultural beliefs and myths (e.g. "come what may!").	Lead discussions moving from symptoms → systems → worldviews → metaphors. Design interventions addressing all layers.	Create a safe space for reflection; document insights visibly to show connections.
Factor analysis	Make sense of complex survey data by uncovering key underlying factors.	Identified two critical success factors: governance and community engagement, and innovation and economic development, which shaped the strategic framework.	Use surveys to gather perceptions, analyse to prioritise resources and actions.	Partner with universities/statisticians. Explain findings in simple terms and link them to action plans.

Designing Cities with People, Not Just for Them: The Power of Collaborative Urban Futures

Archangelo D Alegado, Rec E Eguia and Alan B Alejandrino
ISBN 978-1-83742-369-9
https://doi.org/10.1108/978-1-83742-366-820261018

CONCLUSION: CHOOSING OUR CITY'S FUTURE

"The future is not something to be endured — it is something to be chosen, designed and built. But only if we act together."

THE STAKES OF INACTION

The journey of this research makes clear that the greatest risk facing Digos, and cities like it, is not change itself, but inaction.

Failing to act perpetuates the very conditions that have hindered Digos in the past: fragmented governance, low trust, tokenistic participation and development that benefits a few, while excluding many. These patterns have real consequences: citizens disengage, opportunities are missed, and investments, both social and economic, stagnate.

The scenarios developed in the research paint this vividly: a future where Digos becomes fragmented or efficient but disconnected is not merely undesirable–it is preventable. The question is whether leaders, planners and citizens will allow themselves to drift toward these futures by default, or choose instead to work intentionally toward a shared vision, shared prosperity scenario–the preferred future identified through collaborative foresight.

WHY BUILDING TOMORROW TOGETHER STARTS TODAY

The research findings show that the future is not determined solely by external forces, but by the collective choices made here and now. These choices begin with leadership that catalyses change rather than commands it, with communities that co-create rather than consent passively, and with institutions that innovate and adapt rather than rigidly defend the status quo.

Building tomorrow together does not require waiting for perfect conditions. It starts with small but deliberate steps: creating spaces for authentic dialogue, adopting inclusive tools and frameworks, and institutionalising the lessons learned from Digos City's journey so far.

The Collaborative Foresight Framework–rooted in futures literacy, participatory governance and locally attuned strategies–offers a practical path forward. But even the strongest framework cannot succeed without the willingness of leaders and citizens alike to embrace it, own it and live it out.

A CALL TO ACTION FOR PLANNERS, LEADERS AND COMMUNITIES

The evidence from this research calls on everyone who cares about their city to act–and to act together.

For planners and engineers, this means moving beyond technical fixes to understand and incorporate the aspirations, fears and stories of the people they serve. For local leaders, it means facilitating rather than dictating; cultivating trust and demonstrating integrity at every step. For communities and citizens, it means seeing themselves not as passive beneficiaries but as rightful co-creators of their city's future–showing up, speaking up and holding institutions accountable.

The future of Digos–or of any city–is neither guaranteed nor fixed. It is a choice. It is shaped by the courage to confront uncomfortable truths, the imagination to envision better, and the discipline to act with intention.

This book, rooted in the lived experience of Digos and informed by its people, is an invitation to design cities not just for people, but with them. To choose futures that reflect our highest aspirations rather than our deepest fears. And to begin, today, the quiet, deliberate work of building tomorrow together.

ABOUT THE AUTHORS

Archangelo D Alegado PhD MPSA MTSS MEP is an urban development practitioner, educator and researcher dedicated to shaping cities that are inclusive, resilient and designed with people–not just for them.

He holds a Doctor of Philosophy in Urban and Regional Development Administration, a culmination of years of research on collaborative foresight and participatory governance in cities, which forms the foundation of this book. He also earned master's degrees in Engineering (Civil Engineering) from the University of Southeastern Philippines; in Terrorism and Security Studies from Charles Sturt University, Australia; and in Public Safety Administration from the Philippine Public Safety College in Manila, reflecting his unique interdisciplinary expertise bridging technical, social, governance and security dimensions of urban development.

Beyond academia, he has managed multi-million-peso construction projects, revitalised public parks and heritage sites, and advised senior police and public safety officers as a research mentor at the National Police College, Davao City, Philippines.

He has also contributed to international and national scholarship, publishing on urban foresight and public safety. His passion lies in creating cities that are inclusive, resilient and reflective of the aspirations of their communities.

As an academic leader, he serves as Dean of the College of Engineering at Cor Jesu College, Digos City, Philippines. In this role, he has championed curriculum and teaching innovation, faculty development and institutional accreditation, while mentoring students and professionals in research and leadership.

Rec E Eguia PhD MPSA MM is a distinguished academic leader, researcher and practitioner in development management and public administration. He is a Professor VI at the University of Southeastern Philippines (USeP), where he has also served as Dean of the College of Development Management and Dean of Advanced Studies. Currently, he is the President of the USeP Faculty Federation and sits as a member of the Board of Regents.

An accomplished scholar, Dr Eguia holds a PhD in Development Administration and has completed executive programs at the Massachusetts Institute of Technology (MIT) under the Fulbright Hubert H Humphrey Fellowship and the Special Program on Urban and Regional Studies (SPURS). He is also a graduate of the Senior Executive Class of the Development Academy of the Philippines, where he was recognised with multiple awards for leadership and outstanding capstone work.

Dr. Eguia has worked widely on futures thinking, public safety, governance and development policy, and has been recognised with national and international best paper awards. Beyond academia, he has served as consultant and evaluator for projects funded by the World Bank, United Nations Development Programme (UNDP), Australian Agency for International Development (AUSAID), and the Department of Agriculture. His career reflects a lifelong commitment to scholarship, innovation and public service.

Alan B Alejandrino PhD MPSA MATM is a licensed Civil Engineer specialising in construction engineering and management, with a strong focus on sustainable and resilient infrastructure systems. He is an Associate Professor I at the University of Southeastern Philippines (USeP), where he teaches across both undergraduate and graduate programs and integrates engineering expertise with development management perspectives.

His research spans public safety and security, rural and urban development, and water management systems, addressing pressing issues of sustainable water resource use, flood risk mitigation and climate-resilient infrastructure planning. Dr. Alejandrino has presented his work at international conferences, including studies on water security governance, leadership competency models for public safety officers and contextual intelligence in policy implementation.

Beyond teaching and research, he is actively engaged in extension programs, spearheading community-based initiatives that strengthen local capacities in governance, infrastructure development and environmental sustainability. His contributions ensure that academic outputs translate into tangible benefits for communities.

Dr. Alejandrino is a member of several professional organisations, including the Philippine Institute of Civil Engineers (PICE), the Structural Engineers Association of Davao (SEAD) and the Philippine Association for the Advancement of Science and Technology (PhilAAST). His career reflects a commitment to bridging engineering innovation, research and public service for nation-building.

REFERENCES

The following works informed the ideas, methods and case studies discussed in this book. Readers interested in further exploring the foundations of collaborative urban futures may find them useful.

AARP (2015) *Intergenerational housing in Lyon, France.* AARP Network of Age-Friendly Communities, 8 Domains of Livability. https://www.aarp.org/livable-communities/network-age-friendly-communities/info-2015/domain-3-intergenerational-housing-lyon-france.html (accessed 21/09/2025).

AIDC (Alternative Information & Development Centre) (n.d.) *Amadiba Crisis Committee (ACC).* https://aidc.org.za/partners/amadiba-crisis-comittee/ (accessed 21/09/2025).

AIPH (International Association of Horticultural Producers) (2018) Green City case study: Lyon, France – Le Champ. https://aiph.org/green-city-case-studies/lyon-france-le-champ/ (accessed 21/09/2025).

Ajuntament de Barcelona (2014) *Barcelona Smart City Strategy.* Barcelona City Council, Barcelona, Spain.

Ajuntament de Barcelona (2016) *Barcelona Digital City Plan 2017–2020.* Barcelona City Council, Barcelona, Spain.

Amadiba Eco Adventures (n.d.) Home. https://amadiba.co.za/ (accessed 21/09/2025).

Anttiroiko A-V (2016) City-as-a-platform: the rise of participatory innovation platforms in Finnish Cities. *Sustainability* 8(9): 922.

Arnstein SR (1969) A ladder of citizen participation, *Journal of the American Planning Association* 35(4): 216–224, 10.1080/01944366908977225.

Bailey N (2020) The right to the city: evaluating the changing role of community participation in urban planning in England. In *The Routledge Handbook of Henri Lefebvre, the City and Urban Society* (Leary-Owhin ME and McCarthy JP (eds)). Routledge, London and New York, pp. 422–431.

Bakıcı T, Almirall E and Wareham J (2013) A smart city initiative: the case of Barcelona. *Journal of the Knowledge Economy* 4(2): 135–148.

Bason C (2018) *Leading Public Sector Innovation: Co-creating for a Better Society*, 2nd edn. Policy Press, Bristol, UK.

Björklund F and Kronqvist J (2020) Participatory budgeting in Helsinki: design choices and democratic practices. *Scandinavian Journal of Public Administration* 24(3): 47–67.

Bolton R (2005) Habermas's theory of communicative action and the theory of social capital. *Association of American Geographers Annual Meeting Proceedings* 97(3): 1–15.

Capel H (2009) The transformation of Barcelona's industrial districts: Poblenou and 22@. *Urban Research & Practice* 2(3): 302–319.

Centre for Environmental Rights (2018) Baleni and Others v Minister of Mineral Resources and Others. https://cer.org.za (accessed 21/09/2025).

City of Helsinki (2017) *Helsinki City Strategy 2017–2021: The Most Functional City in the World.*

City of Helsinki (2018) *Carbon Neutral Helsinki 2035 Action Plan.*

Clark DA (2005) *The Capability Approach: Its Development, Critiques and Recent Advances.* Global Poverty Research Group Working Paper Series, GPRG-WPS-032.

Co-VAL (2019) Erasme Living Lab, Lyon Métropole, France: 20 years of public innovations. Co-creation of Value in European Public Services. https://www.co-val.eu/case-studies/blog/project/erasme-living-lab-lyon-metropole-france-20-years-of-public-innovations/(accessed 21/09/2025).

Darchen S (2019) Contextual and external factors enabling planning innovations in a regeneration context: the Lyon Confluence project, France. *International Planning Studies* **24(1)**: 52–67, 10.1080/13563475.2018.1546359.

GovTech Singapore (2018) *Transforming the Singapore Government through Technology*. GovTech Singapore, Singapore.

Habermas J (1984) *Theorie des kommunikativen Handelns.* English (translated by Thomas McCarthy). Beacon Press, Boston, MA, USA.

ICJ (International Commission of Jurists) (2018) South Africa: ICJ welcomes landmark judgment on free and informed consent of communities prior to the award of mining rights, 22 November. https://www.icj.org (accessed 21/09/2025).

IMDA (Infocomm Media Development Authority) (2018) *Digital Readiness Blueprint*. Singapore Government, Singapore.

Inayatullah S (2004) The causal layered analysis (CLA) reader. *Theory and Case Studies of an Integrative and Transformative Methodology* **1**: 1–52.

Kim AM (2011) The mixed-use redevelopment of Seoul: the politics of policy design. *Cities* **28(2)**: 113–118.

Lund DH (2018) Co-creation in urban governance: from inclusion to innovation. *Scandinavian Journal of Public Administration* **22(2)**: 3–17.

Mail & Guardian (2018) High Court rules in favour of Xolobeni community in historic mining rights case. https://mg.co.za (accessed 21/09/2025).

March H and Ribera-Fumaz R (2016) Smart contradictions: the politics of making Barcelona a self-sufficient city. *European Urban and Regional Studies* **23(4)**: 816–830.

Miller R (2007) Futures literacy: a hybrid strategic scenario method. *Futures* **39(4)**: 341–362.

Mnwana S (2015) Mining and 'communal' land rights in South Africa: whose land is it anyway? *Review of African Political Economy* **42(146)**: 79–100, 10.1080/03056244.2014.984188.

OECD (2020) *Innovative Citizen Participation and New Democratic Institutions: Catching the Deliberative Wave*. OECD Publishing, 10.1787/339306da-en.

Prime Minister's Office Singapore (2014, November 24) *PM Lee Hsien Loong at the Smart Nation launch*. https://www.pmo.gov.sg/newsroom/transcript-prime-minister-lee-hsien-loongs-speech-smart-nation-launch-24-november/

Printed and bound by CPI Group (UK) Ltd, Croydon, CR0 4YY

15/09/2026

14962488-0002